Lecture Notes in Applied and Computational Mechanics

Volume 22

Series Editors

Prof. Dr.-Ing. Friedrich Pfeiffer
Prof. Dr.-Ing. Peter Wriggers

Lecture Notes in Applied and Computational Mechanics

Edited by F. Pfeiffer and P. Wriggers

Vol. 23: Frémond M., Maceri F. (Eds.)
Mechanical Modelling and Computational Issues in Civil Engineering
400 p. 2005 [3-540-25567-3]

Vol. 22: Chang C.H.
Mechanics of Elastic Structures with Inclined Members: Analysis of Vibration, Buckling and Bending of X-Braced Frames and Conical Shells
190 p. 2004 [3-540-24384-4]

Vol. 21: Hinkelmann R.
Efficient Numerical Methods and Information-Processing Techniques for Modeling Hydro- and Environmental Systems
305 p. 2005 [3-540-24146-9]

Vol. 20: Zohdi T.I., Wriggers P.
Introduction to Computational Micromechanics
196 p. 2005 [3-540-22820-9]

Vol. 19: McCallen R., Browand F., Ross J. (Eds.)
The Aerodynamics of Heavy Vehicles: Trucks, Buses, and Trains
567 p. 2004 [3-540-22088-7]

Vol. 18: Leine, R.I., Nijmeijer, H.
Dynamics and Bifurcations of Non-Smooth Mechanical Systems
236 p. 2004 [3-540-21987-0]

Vol. 17: Hurtado, J.E.
Structural Reliability: Statistical Learning Perspectives
257 p. 2004 [3-540-21963-3]

Vol. 16: Kienzler R., Altenbach H., Ott I. (Eds.)
Theories of Plates and Shells: Critical Review and New Applications
238 p. 2004 [3-540-20997-2]

Vol. 15: Dyszlewicz, J.
Micropolar Theory of Elasticity
356 p. 2004 [3-540-41835-0]

Vol. 14: Frémond M., Maceri F. (Eds.)
Novel Approaches in Civil Engineering
400 p. 2003 [3-540-41836-9]

Vol. 13: Kolymbas D. (Eds.)
Advanced Mathematical and Computational Geomechanics
315 p. 2003 [3-540-40547-X]

Vol. 12: Wendland W., Efendiev M. (Eds.)
Analysis and Simulation of Multifield Problems
381 p. 2003 [3-540-00696-6]

Vol. 11: Hutter K., Kirchner N. (Eds.)
Dynamic Response of Granular and Porous Materials under Large and Catastrophic Deformations
426 p. 2003 [3-540-00849-7]

Vol. 10: Hutter K., Baaser H. (Eds.)
Deformation and Failure in Metallic Materials
409 p. 2003 [3-540-00848-9]

Vol. 9: Skrzypek J., Ganczarski A.W. (Eds.)
Anisotropic Behaviour of Damaged Materials
366 p. 2003 [3-540-00437-8]

Vol. 8: Kowalski, S.J.
Thermomechanics of Drying Processes
365 p. 2003 [3-540-00412-2]

Vol. 7: Shlyannikov, V.N.
Elastic-Plastic Mixed-Mode Fracture Criteria and Parameters
246 p. 2002 [3-540-44316-9]

Vol. 6: Popp K., Schiehlen W. (Eds.)
System Dynamics and Long-Term Behaviour of Railway Vehicles, Track and Subgrade
488 p. 2002 [3-540-43892-0]

Vol. 5: Duddeck, F.M.E.
Fourier BEM: Generalization of Boundary Element Method by Fourier Transform
181 p. 2002 [3-540-43138-1]

Vol. 4: Yuan, H.
Numerical Assessments of Cracks in Elastic-Plastic Materials
311 p. 2002 [3-540-43336-8]

Vol. 3: Sextro, W.
Dynamical Contact Problems with Friction: Models, Experiments and Applications
159 p. 2002 [3-540-43023-7]

Vol. 2: Schanz, M.
Wave Propagation in Viscoelastic and Poroelastic Continua
170 p. 2001 [3-540-41632-3]

Vol. 1: Glocker, C.
Set-Valued Force Laws: Dynamics of Non-Smooth Systems
222 p. 2001 [3-540-41436-3]

Mechanics of Elastic Structures with Inclined Members

Analysis of Vibration, Buckling and Bending of X-Braced Frames and Conical Shells

Chin Hao Chang

Dr. CHIN HAO CHANG
Professor Emeritus
The University of Alabama
Dept. Aerospace Engineering and Mechanics
P.O.Box 870280
Tuscaloosa, AL 35487-0280
USA

With 115 Figures and 25 Tables

ISSN 1613-7736

ISBN-10 3-540-24384-4 Springer Berlin Heidelberg New York

ISBN-13 978-3-540-24384-7 Springer Berlin Heidelberg New York

Library of Congress Control Number: 2005926501

Springer is a part of Springer Science+Business Media
springeronline.com

Printed in The Netherlands

Typesetting: Data conversion by the author.
Final processing by TechBooks, New Delhi
Cover-Design: design & production GmbH, Heidelberg
Printed on acid-free paper 89/TechBoooks - 5 4 3 2 1 0

Preface

Mechanics of vibration, buckling and bending of elastic structures with inclined members such as x-braced high rise frames and conical shells is studied in this monograph.

There are two basic methods to analyze elastic continuum mechanics of structures: One is based on Newton's law of motion (or equilibrium); the other is the energy method which is by applying variation to Hamilton's principle and Lagrange's equation to obtain the governing equations. Two examples are given, one each for the two approaches, in Chap. 1 as a review. They also serve as basic models for the limited cases of vibrations of inclined members studied in Chap. 2.

There are not many analytic complete solutions of conical shells available in the literature, especially about the stability of conical shells. The buckling loads of conical shells obtained from analytic approximate methods have been much higher than those obtained by the available experiments.

For the buckling of conical shells, there are two unique technical aspects having not been adequately addressed in the literature. One is the upper end condition; the other is the effect of angle change before buckling taking place.

Consider a truncated cone with both ends having rigid bulkheads subjected to a compressive single force acting along the centric line of the cone. This results in the axisymmetric deformation of the cone.

Let the bottom of the cone be fixed in space, the upper end can only be deformed in the vertical direction due to symmetry. This restriction is a constraint in addition to the other geometric and natural boundary conditions to be satisfied by the cone. This constraint is brought into the system by the Lagrange's multiplier as demonstrated by a one-dimensional model of vibrations of inclined bars in Chap. 2.

This model is applied to study the vibrations of frames with inclined members in Chap. 3, to the vibrations of a portal frame with an x-brace in Chap. 4, and to multi-story x-braced building frames in Chap. 5. The results indicate that for high-rise buildings, the top floor will have the most destruction from earthquake due to horizontal ground vibrations.

The effect of shear deformation and rotatory inertia on the vibrations of the inclined bars with end constraint is investigated in Chap. 6.

The vibrations of the frames studied in Chaps. 3, 4 and 5 are all for in-plane vibrations. By combining the transverse and torsional vibrations, out-of-plane vibrations of portal frames without and with x-braces are studied in Chap. 7. The out-of-plane vibration frequencies are compared with those of the in-plane vibrations of the same frames. The results indicate that the basic frequencies of vibrations for out-of-plane frames are lower than those of in-plane frames as expected.

A nonlinear large deflection theory for buckling of the one-dimension model with end constraint subjected to a vertical compressive force at the top end is presented in Chap. 8. The exact solutions for the inclined columns with different end conditions are obtained. It is shown that for simply supported inclined columns, the effect of angle change is significant.

A two-member truss with three hinges is used for the navigation mitered lock gates. The truss is subjected to an uniform hydraulic lateral normal load. Because of symmetry, the truss may be analyzed by using one member which is identical to the inclined bar studied in Chap. 8. The only difference is that now the bar is subjected to uniformly distributed lateral load instead of a single force acted at one end of it. Thus, an inclined member with end constraint subjected to lateral bending load is investigated in Chap. 9.

The wall of a conical shell has an angle of inclination with the vertical line. There is a fundamental difference between the buckling of a column, a plate, or a cylindrical shell and the buckling of conical shells. For instance, a straight column subjected to an axial compressive load has axial deformation, but the configuration does not change; a straight column remains straight, theoretically, until buckling takes place. For a conical shell, however, the wall starts to bend as soon as the load being applied on; the angle of the cone starts to change which will affect the buckling strength of the cone. This effect of the angle change before buckling has not been previously accounted for in the literature for the study of the buckling of conical shells. These effects are significant as demonstrated by an inclined column in Chap. 8, for the mitered gate girder in Chap. 9, and by a conical shell in Chap. 11.

The stability of a conical shell is a nonlinear problem, the edge-zone or boundary-layer technique has been widely used in the large deformation of nonlinear problems for plates as well as for shells. However, there are few experiments to verify these results in the literature. Before applying this technique to the buckling of conical shells, the boundary-layer technique is verified by an experiment to the solutions of a rectangular plate buckling. It is found in Chap. 10 that the solutions obtained by the boundary-layer method agree with the experimental results very well.

In Chap. 11, the buckling of a conical shell subjected to a single compressive force acting along the central line of the cone is studied. Using the variation method, a set of nonlinear equations has been derived. The solutions are separated into two parts: membrane and bending. The bending part is solved by the boundary-layer technique. The two parts are coupled by the lateral displacements.

In the study of conical shell buckling, both the end constraint and the angle change effects are taken into consideration. A satisfactory agreement between the present analytic results and the available experiments in the literature is presented.

Conical shells with linearly varying thickness subjected to normal loads are investigated in Chap. 12. Asymptotic solutions in closed forms are obtained. Asymptotic solutions for conical shells of constant thickness are also presented in Chap. 13. Numerical examples are given in both chapters.

Chapter 14 covers the membrane vibrations of conical shells in both meridional and circumferential directions. In both cases, frequency equations for cones with various boundary conditions are presented.

This monograph is oriented for problem solving. No detailed derivations of basic equations are given for some topics like plates and shells. The monograph may be used as text book for one semester of special study of structural mechanics for graduate students who have at least one year graduate study in mechanics. It may also be used as reference book for engineers and researchers in the fields of engineering mechanics, civil engineering, mechanical engineering and aerospace engineering.

Each chapter of this monograph, except Chap. 1, represents one or two previously published papers in technical journals by the author or with his former graduate students. Most papers were published in the United States of America, where the English system is still allowed. Therefore, the English system was used in a few papers.

These researches were supported by National Science Foundation, NASA George C. Marshall Space Flight Center and the Faculty Fellowships of the Graduate School at the University of Alabama.

Acknowledgements are due to the following former graduate students for their assistance in the programming of the numerical computations involved: H. Y. Chu, P. D. Schmitz, P. Y. Wang, Y. C. Juan, L. Katz, M. D. Cherng and G. S Chang. Acknowledgements are also due to my son, Yu Yang, for helping me to set up the computer system and assisting me in the techniques required to convert the papers to the present book, and finally to my wife, Kathy, for reading over the manuscript.

June 2005 *Chin Hao Chang*

Contents

1

Two Methods for Analysis of Continuum Mechanics of Elastic Structures

There are two basic methods to analyze the continuum elastic structural mechanics: One is by applying Newton's law of motion to a free body element of a structure; the other is by the energy method or calculus of variations [23]*. Two examples of vibration of a bar are used in this chapter to illustrate these two methods as a review.

1.1 Longitude Vibration of a Bar

1.1.1 Formulation – By Equation of Motion

Consider the bar shown in Fig. 1.1(a) as an example [47]. The bar is under longitudinal or axial vibration of $u(x,t)$, where u is the axial displacement; x is the distance from the left-hand end; and t is time. Consider the element of the bar shown in Fig. 1.1(b), in which S is the resultant force of internal stress, σ, acting on the cross-sectional area A of the element which is with acceleration, a. It is assumed that during vibration, the plane cross-section remains plane. The mass of the element with density ρ is $A\rho\, dx$. The law of motion is

$$F = ma$$

where $F =$ force and $m =$ mass. Since the strain $\varepsilon = \partial u/\partial x$, by Hooke's law, the axial force

$$S = A\sigma = AE\varepsilon = AE\;\partial u/\partial x$$

in which $E =$ modulus of elasticity. Substituting this expression into the last equation yields the equation of motion for the system

$$\partial^2 u/\partial t^2 = a^2 \partial^2 u/\partial x^2 \tag{1.1}$$

where

$$a^2 = E/\rho \tag{1.2}$$

* The numbers in brackets designate references listed at the end of the book.

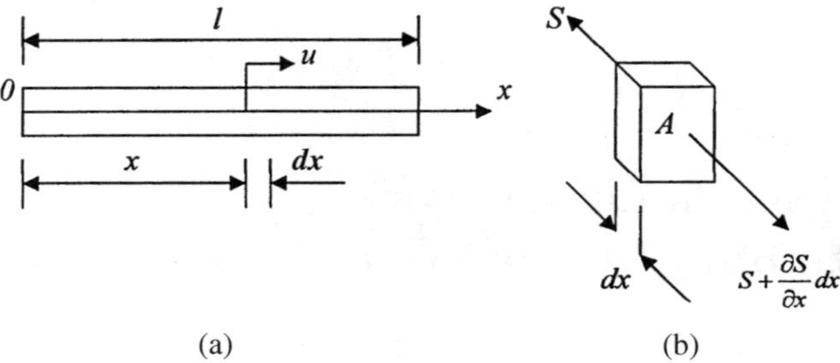

Fig. 1.1. A bar under axial vibration

1.1.2 Solutions

The partial differential (1.1) is solved by the method of separation of variables. Let the displacement function be

$$u(x,t) = X(x)\,T(t) \tag{1.3}$$

Substituting this function into (1.1) and dividing the resulted equation by XT, one has

$$\frac{\ddot{T}}{T} = a^2\frac{X''}{X} = -p^2 \tag{1.4}$$

in which a prime indicates derivative with respective to x, a dot with respective to t, and p is a constant known as eigenvalue. Equation (1.4) may be separated into two equations:

$$a^2X'' + p^2X = 0 \tag{1.5}$$

and

$$\ddot{T} + p^2T = 0 \tag{1.6}$$

The solutions of these two equations are, respectively,

$$X(x) = A\sin(px/a) + B\cos(px/a) \tag{1.7}$$

$$T(t) = C\sin(pt) + D\cos(pt) \tag{1.8}$$

where p is now known as the circular frequency. Equation (1.7) is called the normal function, defining the shape of the natural mode of the vibration. The constants A, B, C, and D are constants of integration to be determined by boundary and initial conditions.

The bar has two kinds of boundary conditions: fixed and free. For fixed ends:

$$X\,(0) = 0 \quad \text{and} \quad X\,(l) = 0 \tag{1.9a,b}$$

and for free ends

$$X'\,(0) = 0 \quad \text{and} \quad X'\,(l) = 0 \tag{1.10a,b}$$

There are two initial conditions:

$$T(0) = g(x) \text{ or } = 0 \quad \text{and} \quad \dot{T}(0) = h(x) \text{ or } = 0 \tag{1.11a,b}$$

where $g(x)$ and $h(x)$ are two prescribed functions.

1.1.3 An Example – A Fixed-Free Bar

Take the bar shown in Fig. 1.1 with the left end fixed and the other end free as an example. The bar has the following two boundary conditions:

$$X(0) = 0 \quad \text{and} \quad X'(l) = 0 \tag{1.12a,b}$$

With satisfying conditions of (1.12a,b), (1.7) requires

$$B = 0 \quad \text{and} \quad \cos(p_n l/a) = 0 \tag{1.13a,b}$$

Equation (1.13b) is known as frequency equation to determine the frequency, p_n. Equation (1.13b) is satisfied if

$$p_n \, l/a = n\pi/2 \qquad n = 1, 3, 5, \ldots \tag{1.14}$$

Let

$$k_n = p_n/a \quad \text{and} \quad \eta = x/l \tag{1.15a,b}$$

Then

$$X_n(\eta) = \sin(k_n \eta) \tag{1.16}$$

The formal solution of (1.3) has the following form

$$\begin{aligned} u(\eta, t) &= \sum_{n=1,3,\ldots} \sin(k_n \, \eta) \left[C_n \sin(p_n \, t) + D_n \cos(p_n \, t) \right] \\ &= \sum_{n=1,3,\ldots} X_n(\eta) \, T_n(t) \end{aligned} \tag{1.17}$$

in which constant A has been combined into C_n and D_n.

1.1.4 Orthogonality of Normal Functions – Clebsch's Theorem [34]

Consider the normal functions X_m and X_n, $m \neq n$, which satisfy (1.5) so that

$$X_m'' = -k_m^2 X_m \tag{1.18a}$$

$$X_n'' = -k_n^2 X_n \tag{1.18b}$$

in which a prime indicates derivative with respect to η. Multiplying (1.18a) by X_n and (1.18b) by X_m, subtracting the resulted equations one from the other, and then integrating by parts, one has

$$\left(k_n^2 - k_m^2\right) \int_1^0 \left(X_m X_n + X_n X_m\right) d\eta = \int_0^1 \left(X_m^{''} X_n^{''} - X_n^{''} X_m\right) d\eta$$

$$= \left[X_m^{'} X_n - X_n^{'} X_m\right]_0^1 - \int_0^1 \left(X_m^{'} X_n^{'} - X_n^{'} X_m^{'}\right) d\eta = 0 \qquad (1.19)$$

The terms in the brackets vanish because of boundary conditions (1.9 and 1.10). Thus

$$\int_0^1 X_m X_n d\eta = 0 \qquad \text{for } k_m \neq k_n \text{ or } m \neq n \qquad (1.20)$$

This relation is known as the orthogonality of the normal functions. One has

$$\int_0^1 X_n^2 d\eta = I_n \qquad (1.21)$$

1.1.5 Transient and Forced Vibrations

If the bar is subjected to an axial force $\Phi(\eta) \sin \Omega t$ per unit length along the bar, the equation of motion (1.1) assumes the form:

$$\ddot{U} - a^2 U'' = \Phi(\eta) \sin \Omega t \qquad (1.22)$$

in which $U(\eta, t)$ is axial displacement with the following initial conditions:

$$U(\eta, 0) = G(\eta) \qquad \text{and} \qquad \dot{U}(\eta, 0) = H(\eta) \qquad (1.23\text{a,b})$$

Solution of (1.22) may be divided into two parts: homogenous and particular

$$U = U_H + U_P \qquad (1.24)$$

The homogenous solution $U_H = u$ satisfies (1.1 to 1.21) except (1.11) which are to be fulfilled by the determination of C_n and D_n in the following normal function expansion.

From (1.17 and 1.23a)

$$U_H(\eta, 0) = \sum_n D_n X_n = G(\eta) \qquad (1.25)$$

Multiplying both sides of the above equation by X_m, integrating the resulted equation, and then applying the orthogonal relations of (1.20 and 1.21), one has

$$D_n = \frac{1}{I_n} \int_0^1 G X_n d\eta \qquad (1.26)$$

Following the same procedure, one obtains

$$C_n = \frac{1}{p_n I_n} \int_0^1 H X_n d\eta \qquad (1.27)$$

This completes the homogeneous solution representing the transient vibration of the system.

For the particular solution, let

$$U_p = \phi(\eta) \sin \Omega t \tag{1.28}$$

from (1.22)

$$-\Omega^2 \phi - a^2 \phi'' = \Phi \tag{1.29}$$

Expanding the prescribed function Φ in the series of the normal functions

$$\Phi = \sum_n E_n X_n(\eta) \tag{1.30}$$

Multiplying both sides by X_m, then integrating the equation from 0 to l, and using (1.20 and 1.21), one has

$$E_n = \frac{1}{I_n} \int_0^1 \Phi X_n d\eta \tag{1.31}$$

Let the seeking function ϕ also be in the series of the normal function, so that

$$\phi(\eta) = \sum_n F_n X_n \tag{1.32}$$

Substituting (1.30 and 1.32) into (1.29) and canceling X_n from both sides of the equation, one has

$$F_n = \frac{E_n}{p_n^2 - \Omega^2} \quad \text{for} \quad p_n \neq \Omega \tag{1.33}$$

The completed solution of (1.29) is

$$U(\eta, t) = \sum_{n=1,3,\ldots} X(\eta) \left[C_n \sin p_n t + D_n \cos p_n t + \frac{E_n}{p_n^2 - \Omega^2} \sin \Omega t \right] \tag{1.34}$$

1.2 Transverse Vibration of a Beam

1.2.1 Formulation – By Energy Method – Hamilton's Principle [23]

Take the beam of length l shown in Fig. 1.2 with cross-sectional area A and density ρ to formulate the problem. Let the central line have a transverse displacement $W(x, t)$ during vibration, where x is the coordinate distance from the left end and t is time.

The kinetic and potential energies of the system are, respectively,

$$T = \frac{A\rho}{2} \int_0^{t_1} \int_0^l (\dot{W})^2 dx dt \tag{1.35a}$$

$$V = \frac{EI}{2} \int_0^{t_1} \int_0^1 (W'')^2 dx dt \tag{1.35b}$$

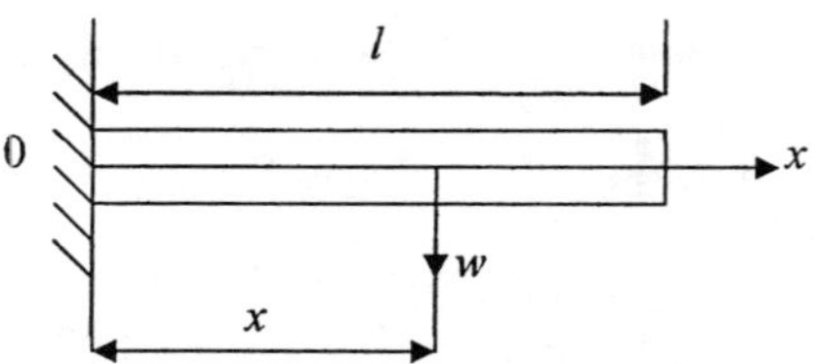

Fig. 1.2. A beam under vibration

in which a dot indicates the derivative with respect to time t and $0 < t < t_1$, where t_1 is an arbitrary time when W or $\dot{W}$ is vanished; a prime indicates derivative with respect to x; and E and I are the modulus of elasticity and moment of inertia of the cross-sectional area, A, respectively. Let

$$\eta = x/l \quad w = W/l \tag{1.36a,b}$$

$$\beta^2 = E/\rho l^2 \quad R^2 = l^2 A/I \tag{1.37a,b}$$

By using the above relations, (1.35) are changed to the following forms:

$$T = \frac{AEl}{2\beta^2}\int_0^{t_1}\int_0^1 (\dot{w})^2 d\eta dt \tag{1.38}$$

$$V = \frac{AEl}{2R^2}\int_0^{t_1}\int_0^1 (w'')^2 d\eta dt \tag{1.39}$$

The Lagrangian function is

$$L = T - V = \frac{AEl}{2}\int_0^{t_1}\int_0^1 \left[\frac{1}{\beta^2}(\dot{w})^2 - \frac{1}{R^2}(w'')^2\right] d\eta dt \tag{1.40}$$

In order to minimize the energy of the system by Hamilton's principle, the variation of the Lagrangian function must vanish. Thus,

$$\delta L = \frac{1}{2}\int_0^{t_1}\int_0^1 \left[\frac{1}{\beta^2}\delta(\dot{w})^2 - \frac{1}{R^2}\delta(w'')^2\right] d\eta dt = 0 \tag{1.41}$$

After performing the variation, integrating the resulted equation by parts, (1.41) yields

$$\frac{1}{\beta^2}\int_0^1 [\dot{w}\delta w]_0^{t_1} d\eta - \frac{1}{R^2}\int_0^{t_1}\left\{[w''\delta w']_0^1 - [w'''\delta w]_0^1\right\} dt$$
$$-\int_0^{t_1}\int_0^1 \delta w\left(\frac{1}{\beta^2}\ddot{w} + \frac{1}{R^2}w''''\right) d\eta dt = 0 \tag{1.42}$$

The vanishing of the third integrand yields the equation of motion for the beam known as the Euler-Bernoulli equation:

$$\frac{1}{\beta^2}\ddot{w} + \frac{1}{R^2}w'''' = 0 \tag{1.43}$$

By letting the terms in the three brackets vanish, the following conditions are obtained:

initial conditions:	$w = 0$	or	$\dot{w} = 0$	at	$t = 0$	and	t_1	(1.44a,b)
boundary conditions:	$w'' = 0$	or	$w' = 0$	at	$\eta = 0$	and	1	(1.45a,b)
	$w''' = 0$	or	$w = 0$	at	$\eta = 0$	and	1	(1.46a,b)

Equations (1.45a and 1.46a) are known as natural boundary conditions and (1.45b and 1.46b) as geometric boundary conditions.

The variation method, thus, not only yields equation of motion but also the appropriate initial and boundary conditions. Furthermore, this method can deal with nonlinear problems, such as the analysis of stability of structures which will be demonstrated for the buckling of a column in Chap. 8, and for a conical shell in Chap. 11. By minimizing the energy of a structural systems, many approximate methods have been developed.

1.2.2 An Example – A Fixed-Free Beam

Take the beam shown in Fig. 1.2 as an example. The beam has the left end fixed and the other end free. Equation (1.43) is also solved by the method of separation of variables. Let

$$w(\eta, t) = X(\eta)\, T(t) \tag{1.47}$$

Following the same procedure as for the axial vibration, the function

$$T(t) = A \sin t + B \cos t \tag{1.48}$$

is introduced, where A and B are constants of integration. The normal function, X, is governed by

$$X'''' - k^4 X = 0 \tag{1.49}$$

where

$$k^4 = p^2 R^2/\beta^2 \tag{1.50}$$

The solution of (1.49) is

$$X(\eta) = C \sin k\eta + D \cos k\eta + E \sinh k\eta + F \cosh k\eta \tag{1.51}$$

which is required to satisfy the following four boundary conditions for the four constants, C, D, E and F.

$$X(0) = 0 \quad \text{hence} \quad F = -D \tag{1.52}$$

$$X'(0) = 0 \quad \text{then} \quad E = -C \tag{1.53}$$

$$X''(1) = k^2 \left[C(-\sin k - \sinh k) - D(\cos k + \cosh k)\right] = 0 \tag{1.54}$$

$$X'''(1) = k^3 \left[C(-\cos k - \cosh k) + D(\sin k - \sinh k)\right] = 0 \tag{1.55}$$

Since $k \neq 0$, (1.54 and 1.55) may be presented in the following matrix form:

$$\begin{bmatrix} \sin k + \sinh k & \cos k + \cosh k \\ \cos k + \cosh k & -\sin k + \sinh k \end{bmatrix} \begin{bmatrix} C \\ D \end{bmatrix} = \begin{bmatrix} 0 \\ 0 \end{bmatrix} \tag{1.56}$$

In order to have the constants C and D to be different from zero or having a non-trivial solution of X, the determinant of the coefficient matrix of (1.56) must vanish. This yields the frequency equation for the beam:

$$1 + \cos k \cosh k = 0 \tag{1.57}$$

which results in

$$k = 1.875,\ 4.694,\ 7.855, \ldots = k_n \quad n = 1, 2, 3, \ldots \tag{1.58}$$

These values are also known as eigenvalues or frequencies of the system. Equation (1.57) and its result, (1.58), are also shown by (2.40 and 2.41) and Fig. (2.7) in Chap. 2, for $\alpha = \pi/2$. From (1.56), the normal function is

$$X_n(\eta) = C_n \left[\sin k_n\eta - \sinh k_n\eta - \frac{\sin k_n + \sinh k_n}{\cos k_n + \cosh k_n} (\cos k_n\eta - \cosh k_n\eta) \right] \tag{1.59}$$

By a similar technique used in the axial vibration of the bar, the normal functions have the following orthogonal relation:

$$\int_0^1 X_m X_n d\eta = 0 \quad \text{for} \quad m \neq n \tag{1.60}$$

One also has

$$I_n = \int_0^1 X_n^2 d\eta \tag{1.61}$$

The normal function expansion method employed for the axial vibration for the bar in the last section can also be followed to obtain the complete solution for the beam.

Exercises

1. Determine the frequencies of the axial vibration of a fixed-fixed bar.
2. Derive the equation of motion and initial and boundary conditions of torsional vibration of a circular shaft by the energy method.
3. Verify the orthogonal relation of (1.60) for the normal functions of beam vibration.
4. Determine the first three frequencies of lateral vibration of a fixed-fixed beam and check the results with those available in Chap. 2.

2

Vibrations of Inclined Bars with End Constraint

Considered are the vibrations of inclined bars with the bottom end fixed in space and the top end having a constraint which requires that this end can move only in the vertical direction.

2.1 One-Dimensional Model of Inclined Members

The axisymmetric vibrations of the frames shown in Fig. 2.1 may be analyzed by a simplified model of an inclined bar with the bottom end fixed in space and the top end moving in a frictionless vertical slot as shown in Fig. 2.2(a). Thus the top end has a constrained condition which requires that the horizontal displacement resulted from the axial displacement, $\bar{U}$, and lateral displacement, $\bar{W}$, at that end must vanish as shown in Fig. 2.2(b):

$$\bar{U}\sin\alpha + \bar{W}\cos\alpha = 0 \quad \text{at } x = l \tag{2.1}$$

in which $\alpha(0 \leq \alpha \leq \pi/2)$ is the angle of inclination of the bar as shown in Figs. 2.1 and 2.2. Condition (2.1) indicates that, in general, the axial displacement, $\bar{U}$, and lateral displacement, $\bar{W}$, are of the same order of magnitude. Thus these two displacements are coupled in the vibrations of the bar.

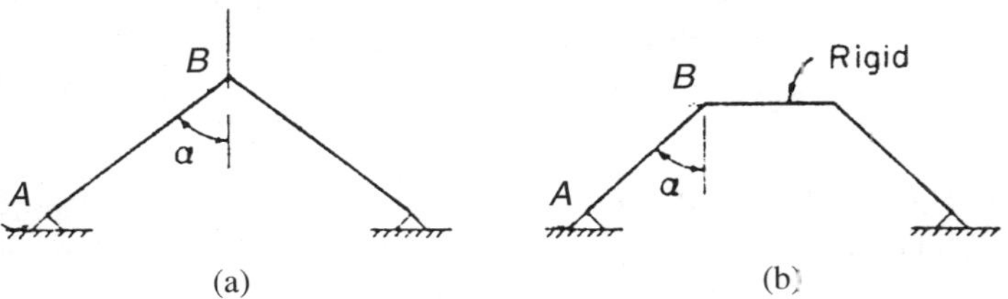

Fig. 2.1. Frames with inclined bars

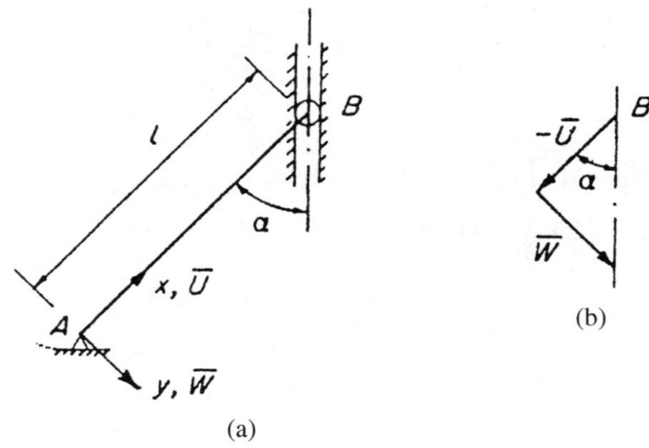

Fig. 2.2. An inclined bar

The coupled vibrations of the inclined bar with different end conditions are studied; the frequencies of small vibrations of such bars are presented; and the orthogonality of the normal functions are formulated in this chapter.

2.2 Formulation for Vibrations of Inclined Bars

Consider the elastic and prismatic inclined bar, AB, of length l shown in Fig. 2.2(a), where x and y are the axial and lateral co-ordinates with the origin located at the centroid of the cross-section at the lower end, A, as shown in Fig. 2.2(a), and let

$$\eta = \frac{x}{l} \quad U = \frac{\bar{U}}{l} \quad \text{and} \quad W = \frac{\bar{W}}{l} \tag{2.2a-c}$$

Then the kinetic energy of the system may be expressed as

$$T = \frac{AEl}{2\beta^2} \int_0^{t_1} \int_0^1 \left(\dot{W}^2 + \dot{U}^2\right) d\eta dt \tag{2.3}$$

in which A is the cross-sectional area; E is the modulus of elasticity; a dot indicates the derivative with respect to time, t; and

$$\beta^2 = Eg/\gamma l^2 \tag{2.4}$$

where g is the gravitational acceleration and γ is the weight of unit volume of the bar. The potential energy of the system for finite vibrations is

$$V = \frac{AEl}{2} \int_0^{t_1} \left[\int_0^1 \left(U_{,\eta} + \frac{1}{2} W_{,\eta}^2\right)^2 d\eta + \frac{1}{R^2} \int_0^1 W_{,\eta\eta}^2 d\eta \right] dt \tag{2.5}$$

in which R is the slenderness ratio and a comma preceded with a subscript represents the derivative with respect to the subscript.

Constrain condition (2.1) may be taken into consideration by using Lagrange's multiplier, λ. Thus, a functional equivalent to Lagrangian function [1, 23] may be formulated as

$$I = T - V - AEl\lambda \int_0^{t_1} (U \sin\alpha + W \cos\alpha)_{\eta=1} \, dt \tag{2.6}$$

When Hamilton's principle is applied to the functional, I, and the virtual displacements δU and δW vanish at the end points of the arbitrary time interval, $0 < t < t_1$, the following two non-linear equations are obtained:

$$\frac{1}{\beta^2}\ddot{U} - \left(U_{,\eta} + \frac{1}{2}W_{,\eta}^2\right)_{,\eta} = 0 \tag{2.7a}$$

$$\frac{1}{\beta^2}\ddot{W} + \left(U_{,\eta} + \frac{1}{2}W_{,\eta}^2\right)_{,\eta} W_{,\eta} + \left(U_{,\eta} + \frac{1}{2}W_{,\eta}^2\right) W_{,\eta\eta} + \frac{1}{R^2}W_{,\eta\eta\eta\eta} = 0 \tag{2.7b}$$

These two nonlinear differential equations can hardly be solved. In view of condition (2.1), one may assume

$$U = \xi U_1 + \xi^2 U_2 + \ldots \qquad W = \xi W_1 + \xi^2 W_2 + \ldots \tag{2.8a,b}$$

where ξ is the amplitude of the vibration, which for a small vibration is a very small number, much less than any dimension of the bar. Therefore, for small vibrations, the terms with second and higher orders of ξ may be neglected. With this simplification, the following two equations for the first order functions U_1 and W_1 are obtained

$$\frac{1}{\beta^2}\ddot{U}_1 - U_{1,\eta\eta} = 0 \qquad \frac{1}{\beta^2}\ddot{W}_1 + \frac{1}{R^2}W_{1,\eta\eta\eta\eta} = 0 \tag{2.9a,b}$$

which are identical to the equations used for vibrations of the bar and beam given in the last chapter, respectively.

The following boundary conditions are obtained from the variation process for bar AB with end A fixed in space:

$$U_1 = 0 \quad \text{at } \eta = 0 \qquad U_{1,\eta} + \lambda \sin\alpha = 0 \quad \text{at } \eta = 1 \tag{2.10a,b}$$

$$W_1 = 0 \quad \text{and} \quad W_{1,\eta} = 0 \quad \text{or} \quad W_{1,\eta\eta} = 0 \quad \text{at } \eta = 0 \tag{2.11a-c}$$

$$W_{1,\eta} = 0 \quad \text{or} \quad W_{1,\eta\eta} = 0 \quad \text{and} \quad W_{1,\eta\eta\eta} - \lambda R^2 \cos\alpha = 0 \quad \text{at } \eta = 1 \tag{2.12a-c}$$

with constraint at $\eta = 1$

$$U_1 \sin\alpha + W_1 \cos\alpha = 0 \tag{2.13}$$

Thus the Lagrange's multiplier, λ, physically is a non-dimensional horizontal reaction at the top end, B. This additional unknown, λ, involved in conditions (2.10b and 2.12c), is compensated by the additional condition (2.13).

2.3 Normal Functions and Frequency Equations

Let

$$U_1(\eta, t) = u(\eta) \sin pt \qquad W_1(\eta, t) = w(\eta) \sin pt \tag{2.14a,b}$$

Satisfying initial conditions $U_1(\eta, 0) = W_1(\eta, 0) = 0$ and substituting these two functions into (2.9), one has

$$u'' + \left(k^4/R^4\right) u = 0 \qquad w'''' - k^4 w = 0 \tag{2.15a,b}$$

in which

$$k = (pR/\beta)^{1/2} \tag{2.16}$$

where p or k is the circular frequency of vibration of this system and a prime indicates the ordinary derivative with respect to η. Conditions (2.10 to 2.13) become, respectively,

$$u = 0 \quad \text{at } \eta = 0 \quad u' + \lambda \sin\alpha = 0 \quad \text{at } \eta = 1 \tag{2.17a,b}$$

$$w = 0 \quad \text{at } \eta = 0 \tag{2.18}$$

$$w' = 0 \quad \text{or} \quad w'' = 0 \quad \text{at } \eta = 0 \tag{2.19a,b}$$

$$w' = 0 \quad \text{or} \quad w'' = 0 \quad \text{at } \eta = 1 \tag{2.20a,b}$$

$$w''' - R^2 \lambda \cos\alpha = 0 \quad \text{at } \eta = 1 \tag{2.21}$$

and

$$u \sin\alpha + w \cos\alpha = 0 \quad \text{at } \eta = 1 \tag{2.22}$$

The solution of (2.15a) satisfying conditions (2.17) is

$$u(\eta) = -\lambda R \sin\alpha \sin(k^2\eta/R)/k^2 \cos(k^2/R) \tag{2.23}$$

The general solution of (2.15b) is

$$w(\eta) = C_1 \sin k\eta + C_2 \cos k\eta + C_3 \sinh k\eta + C_4 \cosh k\eta \tag{2.24}$$

in which C_1, C_2, C_3 and C_4 are four constants of integration to be determined for different types of bars to be studied below.

2.3.1 Inclined Bars with Hinged – "Hinged" Ends

For this type of bar, with satisfaction of conditions (2.18, 2.19b, 2.20b and 2.21), solution (2.24) yields

$$w(\eta) = -\lambda R^2 \cos\alpha \left(\sinh k \sin k\eta + \sin k \sinh k\eta\right)/k^3 \left(\sinh k \cos k - \cosh k \sin k\right) \tag{2.25}$$

On substitution of solutions (2.23 and 2.25) into condition (2.22), for $\lambda \neq 0$, one obtains the frequency equation:

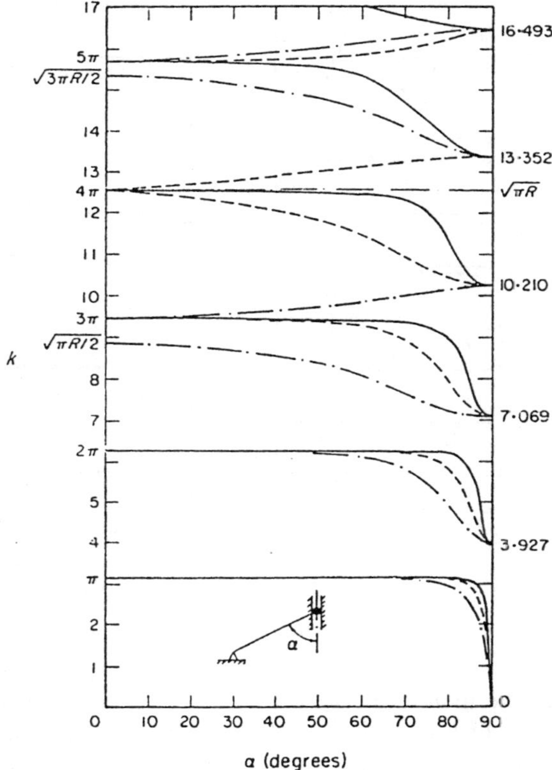

Fig. 2.3. Frequencies of hinged-"hinged" bars —. $R = 200$; $\cdots R = 100$; $- \cdot -$, $R = 50$

$$\tan^2 \alpha - 2(R/k)\cot(k^2/R)(\sin k \sinh k)/(\sin k \cosh k - \cos k \sinh k) = 0 \quad (2.26)$$

With the slenderness ratio, R, as a parameter, the results of this frequency equation are presented in Fig. 2.3 for frequency values, k, versus angle, α. The first five normalized modes of vibration for displacements u and w, for $R = 100$ and $\alpha = 60°$, are depicted in Fig. 2.4.

When $\alpha = 0$, (2.26) calls for either

$$\cos\left(k^2/R\right) = 0 \qquad \text{or} \qquad (2.27)$$

$$\sin k = 0 \qquad (2.28)$$

For $k \neq 0$, these two equations are satisfied by, respectively,

$$k = \sqrt{\pi R/2}, \sqrt{3\pi R/2}, \sqrt{5\pi R/2}, \ldots \qquad \text{and} \qquad (2.29)$$

$$k = \pi, 2\pi, 3\pi, \ldots \qquad (2.30)$$

which are the frequencies of uncoupled axial and lateral vibrations of a hinged-"hinged" bar, respectively.

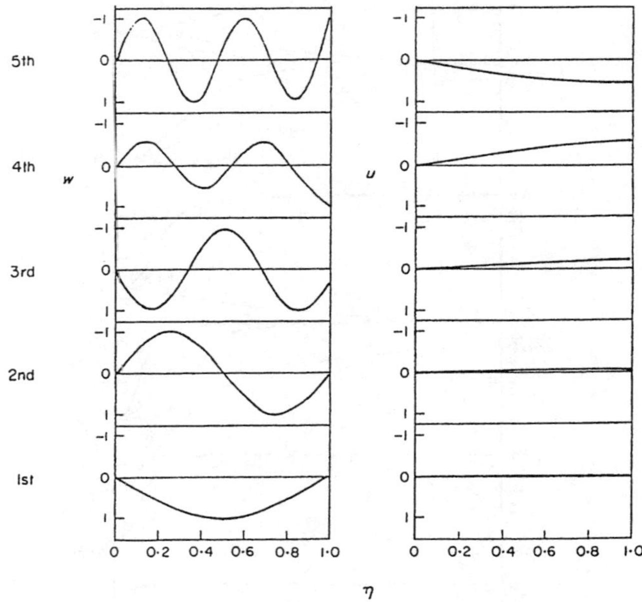

Fig. 2.4. Normal modes of a hinged-"hinged" bar $R = 100$; $R = 50$

For $\alpha = \pi/2$, the first term in (2.26), $\tan^2(\pi/2) = \infty$. The bar becomes a horizontal beam which from (2.18, 2.19b, 2.20b and 2.21) has the following boundary conditions:

$$w(0) = w''(0) = w''(1) = w'''(1) = 0 \tag{2.31}$$

From these conditions, one obtains the frequency equation for the beam as

$$\tan k = \tanh k \tag{2.32}$$

hence,

$$k = 0,\ 3.927,\ 7.069, \ldots \tag{2.33}$$

Equation (2.32) may also be obtained from (2.26) simply by letting the denominator of the second term equal zero or the second term also be infinite. This method to obtain the limiting k values for $\alpha = \pi/2$ is followed by all next three cases.

Let the denominator of (2.26) vanish, another frequency equation is reached for $\pi/2 = 0$:

$$\sin\left(k^2 R\right) = 0 \tag{2.34a}$$

and thus,

$$k = \sqrt{\pi R}, \sqrt{2\pi R}, \sqrt{3\pi R}, \ldots \tag{2.34b}$$

These are frequencies of uncoupled axial vibrations of a fixed-fixed bar.

All these limiting k values, for $\alpha = 0$ and $\pi/2$, are indicated in Fig. 2.3 along the two vertical edges.

2.3.2 Inclined Bars with Bottom Hinged and Top "Clamped"

With the fulfillment of conditions (2.18, 2.19b, 2.20a and 2.21), solution (2.24) becomes

$$w(\eta) = -\lambda R^2 \cos\alpha \left(\cosh k \sin k\eta - \cos k \sinh k\eta\right) / \left(2k^3 \cos k \cosh k\right). \quad (2.35)$$

Condition (2.22), for $\lambda \neq 0$, yields the frequency equation:

$$\tan^2\alpha - (R/2k)\cot\left(k^2/R\right)(\tanh k - \tan k) = 0 \quad (2.36)$$

which, for $\alpha = 0$ and $k \neq 0$, reduces to (2.32). For $\alpha = \pi/2$, (2.36) requires

$$\tan k = \infty \qquad \text{and hence} \quad (2.37a)$$

$$k = \pi/2,\ 3\pi/2,\ 5\pi/2, \ldots \quad (2.37b)$$

which are the frequencies of a guided-hinged beam vibrations. The numerical results of (2.36 and 2.37) are depicted in Fig. 2.5.

The normalized displacements u and w, for $\alpha = 60°$ and $R = 100$, are shown in Fig. 2.6.

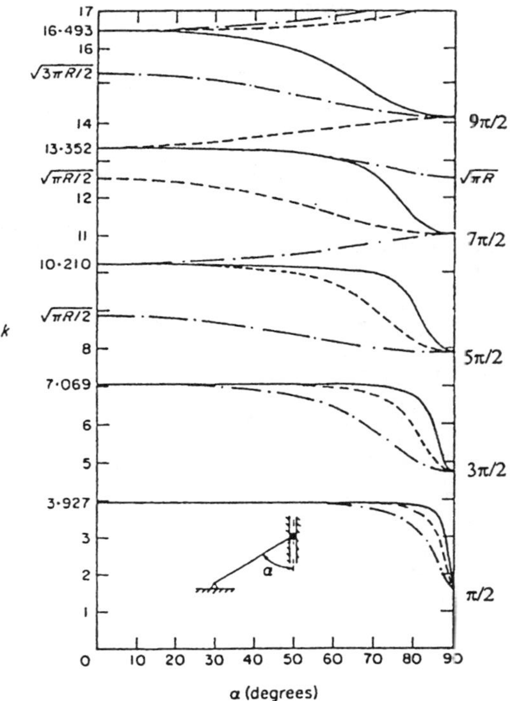

Fig. 2.5. Frequencies of hinged-"clamped" bars —, $R = 200$; ······, $R = 100$; $- \cdot -$, $R = 50$

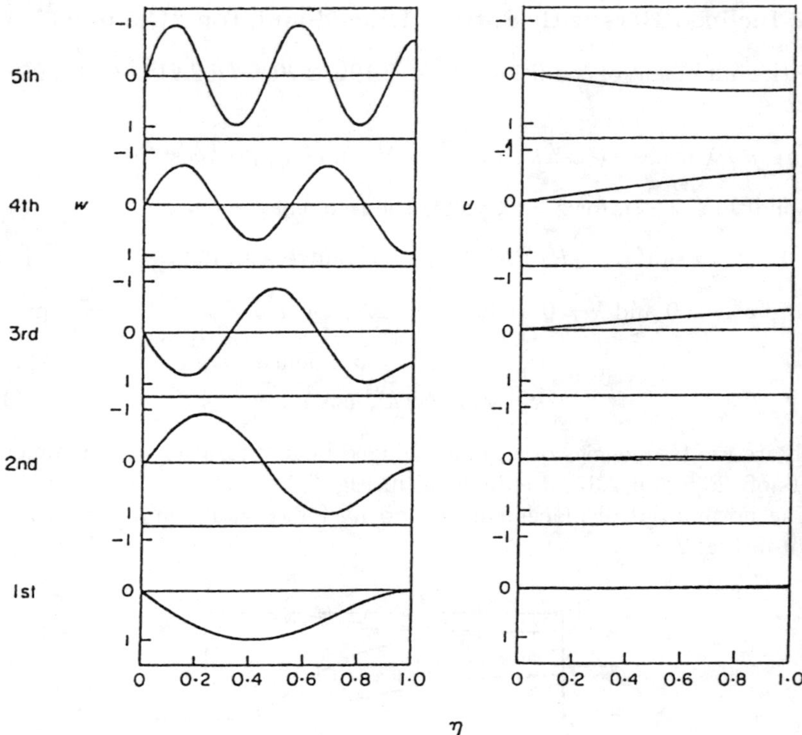

Fig. 2.6. Normal modes of a hinged-"clamped" bar $R = 100$, $\alpha = 60°$

2.3.3 Inclined Bars with Bottom Clamped and Top "Hinged"

Solution (2.24) satisfying (2.18, 2.19a, 2.20b and 2.21) has the following form:

$$\begin{aligned} w(\eta) = &- \left[\lambda R^2 \cos\alpha / 2k^3 \left(1 + \cos k \cosh k\right)\right] \left[(\cos k + \cosh k)\right. \\ &\times \left.(\sin k\eta - \sinh k\eta) - (\sin k + \sinh k)(\cos k\eta - \cosh k\eta)\right] \end{aligned} \quad (2.38)$$

Condition (2.22), for $\lambda \neq 0$, results in the following frequency equation:

$$\tan^2\alpha - (R/k)\cot\left(k^2/R\right)(\sinh k \cos k - \cosh k \sin k) / (1 + \cos k \cosh k) = 0 \quad (2.39)$$

The results of (2.39) are presented in Fig. 2.7.

When $\alpha = 0$, (2.39) also reduces to (2.32), for $k \neq 0$. When $\alpha = \pi/2$, one has

$$1 + \cos k \cosh k = 0 \quad (2.40)$$

from which

$$k = 1.875,\ 4.694,\ 7.855, \ldots \quad (2.41)$$

as given by (1.57 and 1.58) in the last chapter.

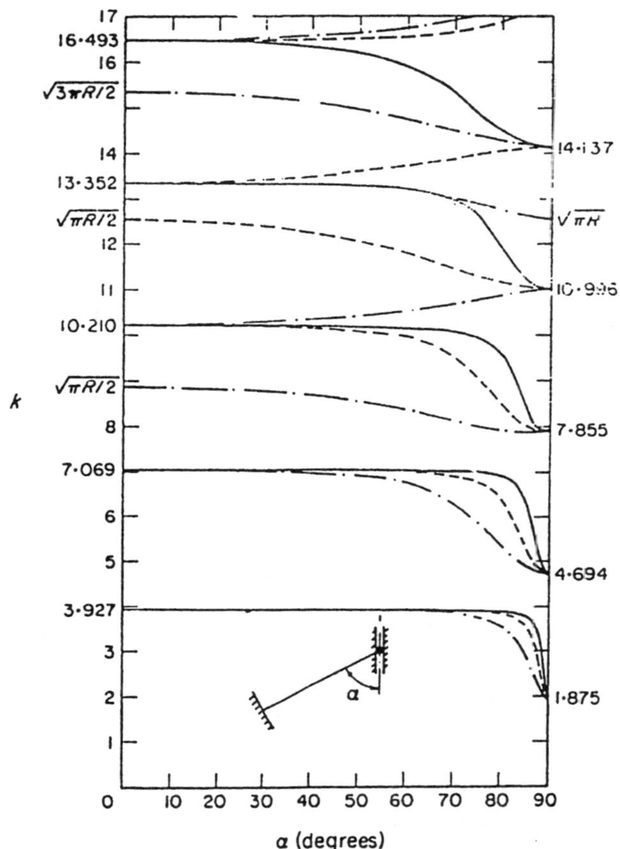

Fig. 2.7. Frequencies of clamped-"hinged" bars —, $R = 200$; ······, $R = 100$; $-\cdot-$, $R = 50$

2.3.4 Inclined Bars with Bottom Clamped and Top Clamped in Rotation

With the satisfaction of (2.18, 2.19a, 2.20a and 2.21), solution (2.24) has the following form:

$$w(\eta) = -\left[\lambda R^2 \cos\alpha / 2 \left(\sin k \cosh k + \cos k \sinh k\right)\right] \left[(\sin k + \sinh k) \times (\sin k\eta - \sinh k\eta) + (\cos k - \cosh k)\,(\cos k\eta - \cosh k\eta)\right] \quad (2.42)$$

The frequency equation obtained from (2.22) is

$$\tan^2\alpha - (R/k)\cot\left(k^2/R\right)(\cos k \cosh k - 1) / (\sin k \cosh k + \cos k \sinh k) = 0 \quad (2.43)$$

The results of this equation are given in Fig. 2.8.

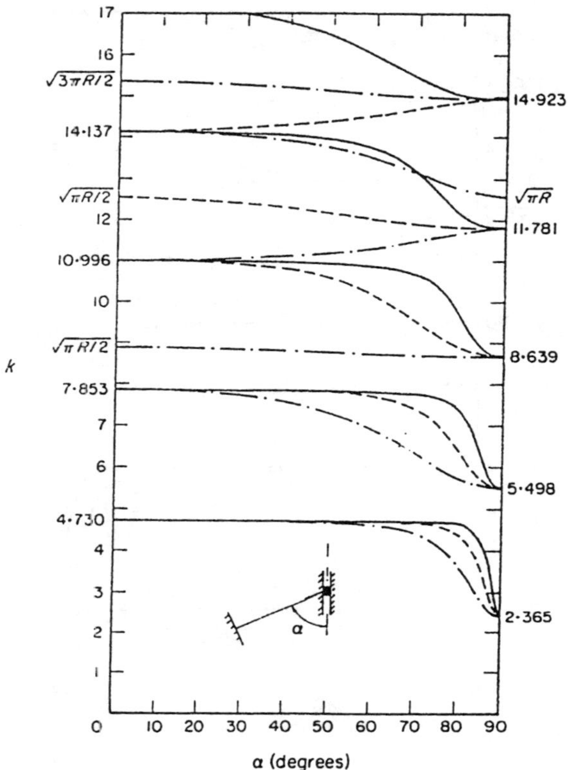

Fig. 2.8. Frequencies of clamped-"clamped" bars —, $R = 200$; ······, $R = 100$; $-\cdot-$, $R = 50$

When $\alpha = 0$, (2.43) yields

$$\cos k \cosh k - 1 = 0 \tag{2.44}$$

from which

$$k = 4.730,\ 7.853,\ 10.996, \ldots \tag{2.45}$$

These are frequencies for the vibrations of a fixed-fixed beam.

When $\alpha = \pi/2$, (2.43) assumes

$$\tan k + \tanh k = 0 \tag{2.46}$$

which results in

$$k = 2.365,\ 5.498,\ 8.639, \ldots \tag{2.47}$$

for a clamped-guided beam.

The normal modes of the last two types of bar vibrations are almost the same as those given in Figs. 2.4 and 2.6, respectively, except that the slopes at $\eta = 0$ vanish. Thus, these figures are not shown here.

It is noted that the slenderness ratio, R, which is not presented in the frequency equations of beam vibrations such as (2.28, 2.32, 2.37a, 2.44 and 2.46) for $\alpha = 0$ and $\pi/2$, plays an important role for $0 < \alpha < \pi/2$, particularly for high modes and small R.

The frequency curves have two alternative groups divided by frequencies of $k = (\pi R/2)^{1/2}$, $(3\pi R/2)^{1/2}$, $(5\pi R/2)^{1/2}, \ldots$, and $(\pi R)^{1/2}$, $(2\pi R)^{1/2}$, $(3\pi R)^{1/2}, \ldots$, as given by (2.29 and 2.34b), respectively.

For instance, it is seen from Figs. 2.3 and 2.4 that when $R = 100$ and $k = (50\pi)^{1/2} = 12.533$, k decreases as α increases below this value, and the end of the lateral displace, w, $(\eta = 1)$ is in tension; while above this value, both k and α increase and the end of the lateral displacement becomes in compression. For smaller R, this pattern changes more often.

The basic modes essentially are of vibrations of a bar with $\alpha = 0$. When α approaches $\pi/2$, however, the basic frequencies drop rather rapidly.

2.4 Orthogonality of the Normal Functions

The mth normal functions satisfy, as shown by (2.15a,b),

$$R^2 u_m'' + k_m^4 u_m = 0 \quad \text{and} \quad w_m'''' - k_m^4 w_m = 0 \tag{2.48a,b}$$

Following the Clebsch's theorem [34],

$$\begin{aligned}
\left(k_n^4 - k_m^4\right) & \int_0^1 \left(w_n w_m + u_n u_m\right) d\eta \\
&= \int_0^1 \left(w_n'''' w_m - R^2 u_n'' u_m\right) d\eta - \int_0^1 \left(w_m'''' w_n - R^2 u_m'' u_n\right) d\eta \\
&= \left[w_n''' w_m - R^2 u_n' u_m - w_m''' w_n + R^2 u_m' u_n\right]_0^1 \\
&\quad - \left[w_n'' w_m' - w_m'' w_n'\right]_0^1 = 0 \,.
\end{aligned} \tag{2.49}$$

In the last step, boundary conditions (2.17 to 2.21) and constraint (2.22) are used. Thus the orthogonal condition of the normal functions reads

$$\int_0^1 \left(w_n w_m + u_n u_m\right) d\eta = 0 \qquad \text{for } n \neq m\,, \tag{2.50}$$

and

$$I_n = \int_0^1 \left(w_n^2 + u_n^2\right) d\eta \tag{2.51}$$

One may use these properties to find the transient and forced vibrations of the system.

Exercises

1. Derive the frequency equation for a hinged-“hinged” beam and check the result with (2.26).
2. Derive frequency (2.44 and 2.46) from a beam vibrations.
3. Show that in (2.49), $[w_n''' w_m - R^2 u_n' u_m - w_m''' w_n + R^2 u_m' u_n] = 0 \quad \text{for } \eta = 1.$

3

Vibrations of Frames with Inclined Members

The model established for the vibration of the inclined bars in the last chapter now is applied to two- and three-bar frames with inclined members.

3.1 Formulation

Consider an elastic plane frame consisting of j prismatic bars. Let $\bar{U}_i\,(t,\eta_i)$ and $\bar{W}_i\,(t,\eta_i)$ be the axial and lateral displacement functions of time, t, and the space coordinate, η_i, measured from one end of the ith bar as shown in Fig. 3.1. The $\bar{U}_i$, $\bar{W}_i$, and η_i are non-dimensionalized with respect to the length l_i.

The equations of motion for $\bar{W}_i$ and $\bar{U}_i$ given by (2.9) are

$$\ddot{\bar{W}}_i + (\beta_i/R_i)^2\,\bar{W}_i'''' = 0 \quad \text{and} \quad \ddot{\bar{U}}_i - \beta^2\bar{U}'' = 0 \tag{3.1a,b}$$

in which a dot indicates the derivative with respect to t and a prime to η_i; R_i is the slenderness ratio

$$R_i^2 = l_i^2 A_i/I_i \quad \text{and} \quad \beta_i^2 = E_i g/\gamma_i l_i^2 \tag{3.2a,b}$$

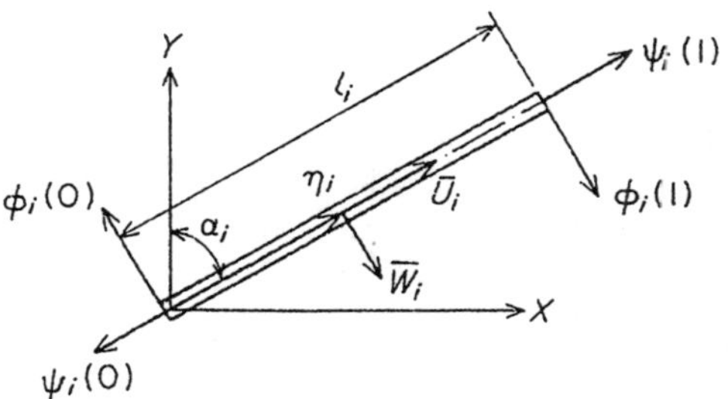

Fig. 3.1. An element of a frame

where E_i is the modulus of elasticity; g is the gravitational acceleration; γ_i is the specific weight; and I_i and A_i are the moment of inertia and cross-sectional area, respectively, of the ith bar. By the method of separation of variables, the solutions of (3.1) may be presented as

$$\bar{W}_i = W_i(\eta_i)\sin pt \qquad \bar{U}_i = U_i(\eta_i)\sin pt \tag{3.3a,b}$$

where

$$W_i(\eta_i) = C_{i1}\sin k_i\eta_i + C_{i2}\cos k_i\eta_i + C_{i3}\sinh k_i\eta_i + C_{i4}\cosh k_i\eta_i \tag{3.4a}$$

$$U_i(\eta_i) = C_{i5}\sin\left(k_i^2\eta_i/R_i\right) + C_{i6}\cos\left(k_i^2\eta_i/R_i\right) \tag{3.4b}$$

in which

$$k_i^2 = pR_i/\beta_i \tag{3.5}$$

where p is the radian frequency of vibration of the system. The constants of integration, C_{is} ($s = 1, 2, \ldots, 6$), are to be determined by the end conditions.

The axial and shearing forces in the bar are, respectively,

$$\psi_i = A_iE_iU_i' \quad \text{and} \quad \phi_i = \frac{A_iE_i}{R_i^2}W_i''' \tag{3.6a,b}$$

Let f_{xi} and f_{yi} be the force components in the directions of the global reference coordinates x and y, respectively. From Fig. 3.1, one obtains

$$f_{xi}(\eta_i) = \psi_i\sin\alpha_i + \phi_i\cos\alpha_i \quad f_{yi}(\eta_i) = \psi_i\cos\alpha_i - \phi_i\sin\alpha_i \tag{3.7,3.8}$$

The x- and y-direction displacement components are

$$\delta_{xi} = (U_i\sin\alpha_i + W_i\cos\alpha_i)\,l_i \quad \delta_{yi} = (U_i\cos\alpha_i - W_i\sin\alpha_i)\,l_i \tag{3.9,3.10}$$

The α_i is the angle of inclination measured from the y-direction as shown in Fig. 3.1.

Without loss of generality, one may consider the three-bar frame shown in Fig. 3.2 as an example. There are $3 \times 6 = 18$ constants of integration to be determined by the following eighteen conditions: nine end conditions:

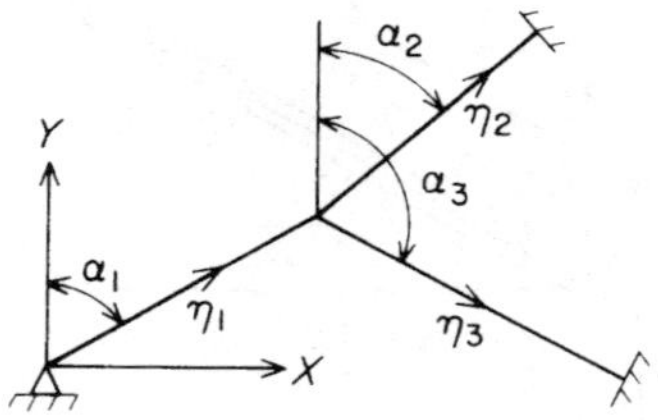

Fig. 3.2. A plane frame

$$W_1(0) = W_2(1) = W_3(1) = 0 \tag{3.11a-c}$$

$$U_1(0) = U_2(1) = U_3(1) = 0 \tag{3.12a-c}$$

$$W_1'(0) \text{ or } W_1''(0) = 0 \quad W_2'(1) \text{ or } W_2''(1) = 0 \quad W_3'(1) \text{ or } W_3''(1) = 0 \tag{3.13a-f}$$

and nine conditions of equilibrium and continuity at the rigidly connected joints:

$$f_{x1}(1) - [f_{x2}(0) + f_{x3}(0)] = 0 \quad f_{y1}(1) - [f_{y2}(0) + f_{y3}(0)] = 0 \tag{3.14,3.15}$$

$$\delta_{x1}(1) - \delta_{x2}(0) = 0 \quad \delta_{\mathrm{x}2}(1) - \delta_{x3}(0) = 0 \tag{3.16a,b}$$

$$\delta_{y1}(1) - \delta_{y2}(0) = 0 \quad \delta_{y2}(1) - \delta_{y3}(0) = 0 \tag{3.17a,b}$$

$$W_1'(1) - W_2'(0) = 0 \quad W_2'(1) - W_3'(0) = 0 \tag{3.18a,b}$$

$$\left(A_1E_1l_1/R_1^2\right)W_1''(1) - \left[\left(A_2E_2l_2/R_2^2\right)W_2''(0) + \left(A_3E_3l_3/R_3^2\right)W_3''(0)\right] = 0 \tag{3.19}$$

If the central joint is hinged, the last three conditions are replaced by

$$W_1''(1) = W_2''(0) = W_3''(0) = 0 \tag{3.20a-c}$$

By using (3.4), the eighteen linear homogeneous equations may be presented in a matrix form

$$[Z]\,[C] = 0 \tag{3.21}$$

in which $[Z]$ is an eighteen by eighteen square coefficient matrix and $[C]$ is a column matrix of eighteen constants, $C_{is}(i = 1, 2, 3$ and $s = 1, 2, \ldots, 6)$. For non-trivial solutions, $C_{is} \neq 0$; hence the determinant of matrix $[Z]$ must vanish from which, the frequencies, p, implicitly included in the parameter k_i in (3.5), can be determined. Then by use of any seventeen of these equations, the corresponding normal functions can be determined with one set of arbitrary constant.

3.2 Vibrations of Two-Bar Frames

Frames made of two identical bars as shown in Fig. 3.3(a) are considered first. For such a frame, there are twelve C's in solutions of (3.4) and $\alpha_1 = \alpha$ and $\alpha_2 = \pi - \alpha$. The system has the following eight conditions:

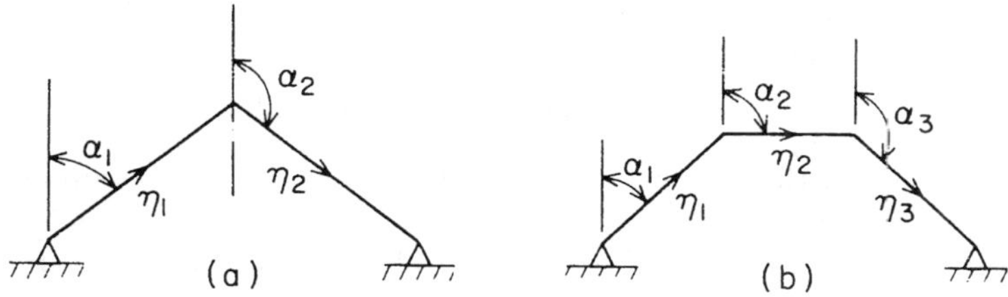

Fig. 3.3. Frames with inclined members

$$W_1(0) = 0 \quad U_1(0) = 0 \tag{3.22a,b}$$

$$W_2(1) = 0 \quad U_2(1) = 0 \tag{3.23a,b}$$

$$\left[U_1' \cos\alpha + W_1''' \sin\alpha / R^2\right](1) - \left[-U_2' \cos\alpha + W_2''' \sin\alpha / R^2\right](0) = 0 \tag{3.24}$$

$$\left[U_1' \sin\alpha - W_1''' \cos\alpha / R^2\right](1) - \left[U_2' \sin\alpha + W_2''' \cos\alpha / R^2\right](0) = 0 \tag{3.25}$$

$$\left[U_1 \sin\alpha + W_1 \cos\alpha\right](1) - \left[U_2 \sin\alpha - W_2 \cos\alpha\right](0) = 0 \tag{3.26}$$

$$\left[U_1 \cos\alpha - W_1 \sin\alpha\right](1) - \left[-U_2 \cos\alpha - W_2 \sin\alpha\right](0) = 0 \tag{3.27}$$

The other four conditions depend on the particular type of the frames under consideration: i.e., whether the ends and the central joint are hinged or rigidly connected.

3.2.1 Frames with All Joints Hinged

For this type of frame, the other four conditions are

$$W_1''(0) = 0 \qquad W_1''(1) = 0 \tag{3.28a,b}$$

$$W_2''(0) = 0 \qquad W_2''(1) = 0 \tag{3.29a,b}$$

With the slenderness ratio, R, as a parameter, the resulting eigenvalues, k, versus the angle, α, are presented in Fig. 3.4. For $R = 50$, the first eight normal modes for $\alpha = 30°$ and the first seven modes for $\alpha = 60°$ are depicted in Figs. 3.5 and 3.6, respectively.

These results indicate that the principal modes of vibrations of the frames may be separated into two groups: symmetric and antisymmetric. For the symmetric modes, the central joint vibrates along the vertical line of symmetry. The frame, thus, may be analyzed by the first bar as in Chap. 2.

In the present approach, however, the bar has a constraint in the horizontal direction:

$$\delta x_1 = 0 \quad \text{or} \quad U_1 \sin\alpha + W_1 \cos\alpha = 0 \quad \text{at } \eta_1 = 1 \tag{3.30}$$

and has no resistance in the vertical direction:

$$f_{y1} = 0 \quad \text{or} \quad U_1' \cos\alpha - \left(W_1'''/R^2\right) \sin\alpha \quad \text{at } \eta_1 = 1 \tag{3.31}$$

The last two conditions combined with the four conditions of (3.22 and 3.28) result in the following frequency equation:

$$\tan^2\alpha - 2\cot\left(k^2/R\right) \frac{R \sin k \sinh k}{k\left(\sin k \cosh k - \cos k \sinh k\right)} = 0 \tag{3.32}$$

which is identical with (3.26) in the last chapter, although the approaches are different. From the comparison of Fig. 2.4 with Fig. 2.3, it is evident that the group of frequency curves represented by (3.32) is for the symmetrical modes. For the antisymmetrical modes, the frequency equation may be obtained by replacing conditions (3.30 and 3.31) with the following two conditions:

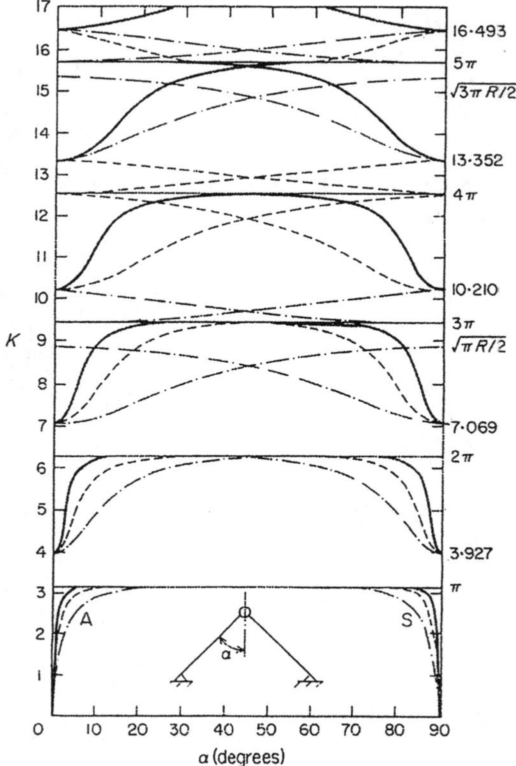

Fig. 3.4. Frequencies of hinged-hinged two-bar frames —, $R = 200$; ······, $R = 100$; $- \cdot -$, $R = 50$

$$\delta_{y1} = 0 \quad \text{or} \quad U_1 \cos\alpha - W_1 \sin\alpha = 0 \text{ at } \eta_1 = 1 \tag{3.33}$$

$$f_{x1} = 0 \quad \text{or} \quad U_1' \sin\alpha + \left(W_1'''/R^2\right)\cos\alpha = 0 \quad \text{at } \eta_1 = 1 \tag{3.34}$$

The resulted frequency equation is

$$\cot^2\alpha - 2\cot\left(k^2/R\right)\frac{R\sin k \sinh k}{k\left(\sin k \cosh k - \cos k \sinh k\right)} = 0 \tag{3.35}$$

Since $\cot\alpha = \tan(\pi/2 - \alpha)$, the curves of (3.35) are the images of those of (3.32) with respect to $\alpha = \pi/4$ as shown in Fig. 3.4.

It is observed that the basic mode is antisymmetric if $\alpha < \pi/4$ as shown by Fig. 3.5 and symmetric if $\alpha > \pi/4$ as shown in Fig. 3.6. In Fig. 3.4, the basic frequency curves of antisymmetric modes are labeled by "A" and of symmetric modes by "S". These two types of modes come up alternatively. However, this alternative pattern breaks up whenever the frequencies pass over the frequencies of axial vibrations of the bar. These are $\sqrt{\pi R/2}$, $\sqrt{\pi R}$, $\sqrt{3\pi R/2}$, ...

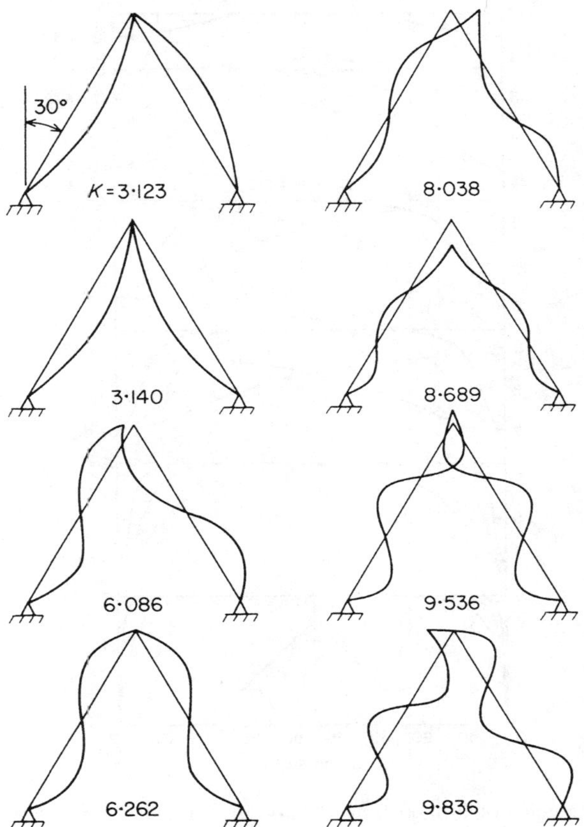

Fig. 3.5. Normal modes of a hinged-hinged two bar frame $R = 50; \alpha = 30°$

(for $R = 50$, $\sqrt{\pi 50/2} = 8.862$). For the frames of $\alpha = 30°$ and $60°$ shown in Figs. 3.5 and 3.6, after $k = 8.689$, the alternative pattern changed.

The first two frequencies, symmetric and antisymmetric, are very close for $10° < \alpha < 80°$. One of them could be easily overlooked in the analysis of a single frame either in experiments or in computations by the trial-and-error method. For instance, the first and second frequencies, $k = 3.123$ and 3.140, respectively, for both $\alpha = 30°$ and $60°$. In fact, the frequencies are identical for both sets. The fact of the antisymmetric mode coming first when $\alpha < \pi/4$ and the symmetrical mode coming first for $\alpha > \pi/4$ may be comprehended by the following observations of the two limited cases.

When $\alpha = 0$, two bars are connected in parallel. Thus, for the first mode, the two bars will bend in the same direction, which forms an antisymmetric mode. On the other hand when $\alpha = \pi/2$, the two bars are connected in series, becoming a continuous beam with a hinge at the center and hence the first mode is symmetrical.

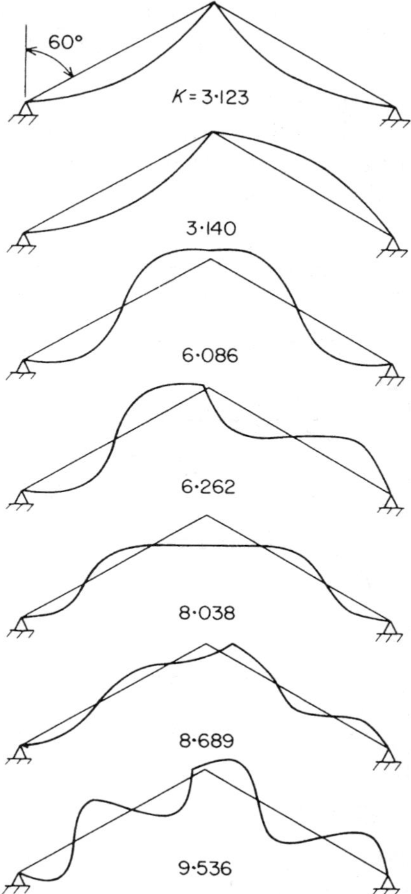

Fig. 3.6. Normal modes of a hinged-hinged two bar frame $R = 50; \alpha = 60°$

3.2.2 Frames with Two Ends Fixed and Central Joint Hinged

This type of frames differ from the preceding ones only in conditions (3.28a and 3.29a) which now are replaced by

$$W_1'(0) = 0 \quad \text{and} \quad W_2'(1) = 0 \tag{3.36a,b}$$

The frequency curves are presented in Fig. 3.7. The frequency equation for the symmetrical modes obtained from (3.22, 3.28b, 3.30, 3.31 and 3.36a) is

$$\tan^2 \alpha - \cot\left(k^2/R\right) \frac{R\left(\sinh k \cos k - \cosh k \sin k\right)}{k\left(1 + \cos k \cosh k\right)} = 0 \tag{3.37}$$

which is identical to (2.39). For the antisymmetric modes, with (3.30 and 3.31) replaced by (3.33 and 3.34), the frequency equation is

$$\cot^2 \alpha - \cot \left(k^2/R\right) \frac{R \left(\sinh k \cos k - \cosh k \sin k\right)}{k \left(1 + \cos k \cosh k\right)} = 0 \qquad (3.38)$$

Thus, the curves of last equation are the images of the curves of (3.37) with respect to $\alpha = \pi/4$ as indicated in Fig. 3.7. The normal modes are similar to those of the preceding type except that the slopes at the two ends will be changed because of conditions (3.36).

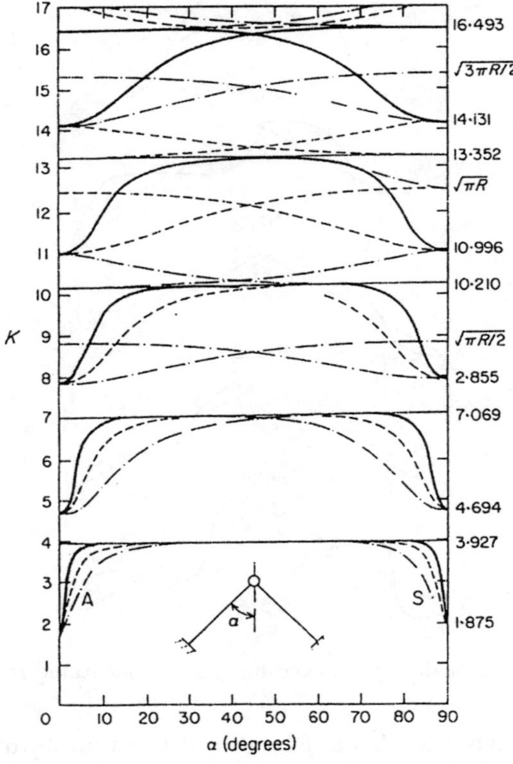

Fig. 3.7. Frequencies of fixed-hinged two-bar frames —, $R = 200$; ·······, $R = 100$; – · –, $R = 50$

3.2.3 Frames with All Joints Rigidly Connected

The four end conditions for this type of frames are

$$W_1'(0) = 0 \qquad W_2'(1) = 0 \qquad (3.39\text{a,b})$$

$$W_1'(1) = W_2'(0) \qquad \text{and} \qquad W_1''(1) = W_2''(0) \qquad (3.40\text{a,b})$$

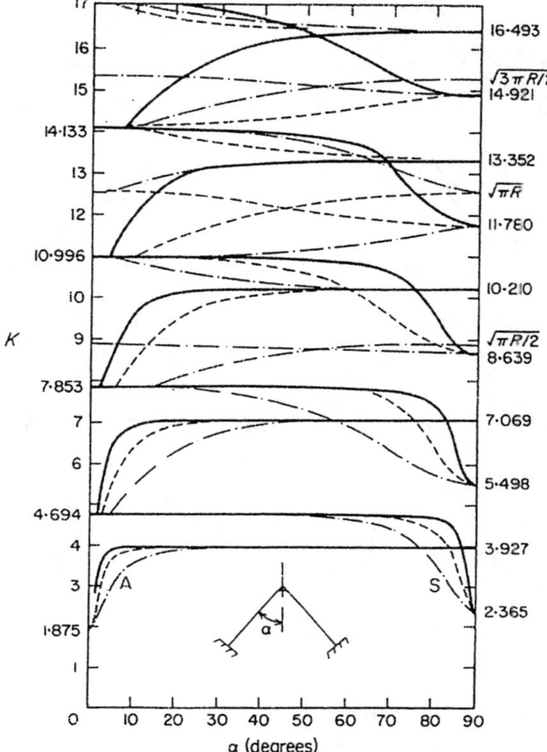

Fig. 3.8. Frequencies of all joints rigidly connected —, $R = 200$; ······. $R = 100$; $- \cdot -$, $R = 50$

From these conditions, one can derive the normal functions and frequency equations. The frequency curves are presented in Fig. 3.8. The frequency equation for symmetric modes derived from the six conditions (3.22, 3.30, 3.31 and 3.39a) and

$$W_1'(1) = 0 \tag{3.41}$$

is

$$\tan^2\alpha - \cot\left(k^2/R\right)\frac{R(\cos k\cosh k - 1)}{k(\sin k\cosh k + \cos k\sinh k)} = 0 \tag{3.42}$$

which is identical with (2.43). The antisymmetric modes can not be analyzed from a single bar, because at the central joint neither $W_1'(1)$ nor $W_1''(1)$ vanishes. Thus there is no alternative way to determine the frequencies of the antisymmetric modes.

For the present type of frames, the basic mode is antisymmetric for most α values. The symmetrical modes come first only when α is close to $\pi/2$.

The normal modes for $R = 100$ and $\alpha = 45°$ are presented in Fig. 3.9. As observed from Fig. 3.8, the first normal mode is antisymmetric. The

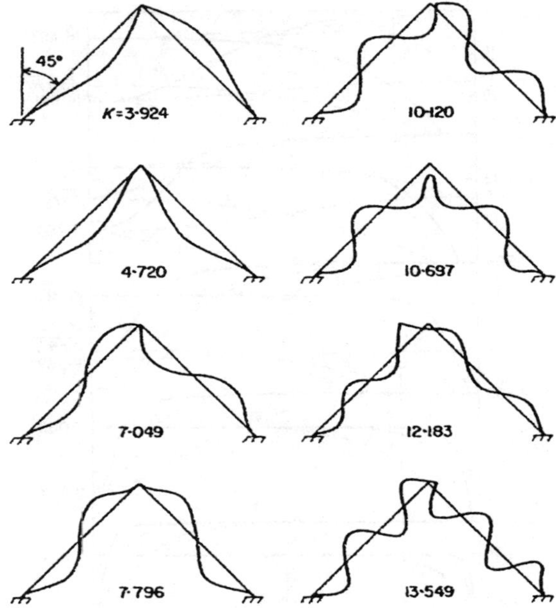

Fig. 3.9. Normal modes of all joints rigidly connected frames $R = 100$; $\alpha = 45^\circ$

alternative pattern of antisymmetric and symmetric modes breaks down as k passed $\sqrt{\pi 100/2} = 12.533$.

3.2.4 Frames with Two Ends Hinged and Central Joint Rigidly Connected

This type of frame is the same as the last one, except that conditions (3.39) now are replaced by

$$W_1''(0) = 0 \quad \text{and} \quad W_2''(1) = 0 \tag{3.43a,b}$$

The resulting frequency curves are presented in Fig. 3.10. For symmetric vibrations, the frequency equation obtained from a single bar with the boundary conditions of (3.22, 3.30, 3.31, 3.41 and 3.43a) is

$$\tan^2\alpha - (R/2k)\cot\left(k^2/R\right)(\tanh k - \tan k) = 0 \tag{3.44}$$

which is identical to (2.36). For the same reason as for the last type, the frequency curves of the antisymmetric modes can not be obtained by a single bar and are not the images of those of symmetrical modes as shown in Fig. 3.10.

For the present type of frames as for the last type, the basic mode is antisymmetric for most α values. The symmetrical modes come first only when α is close to $\pi/2$.

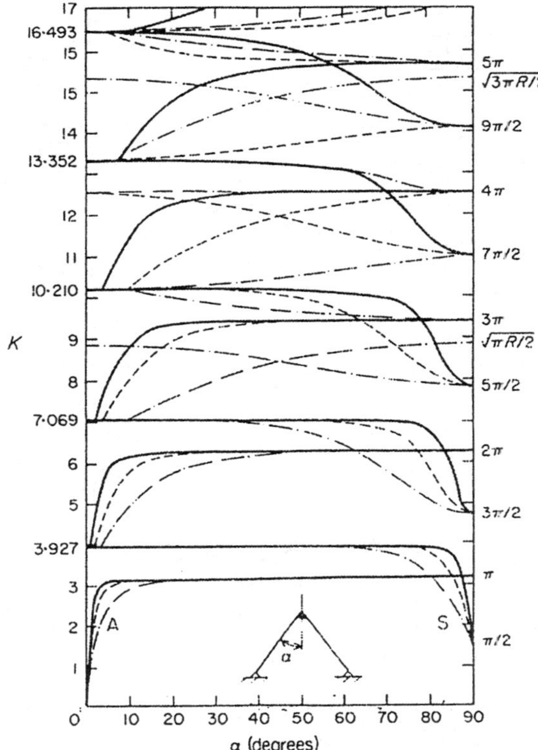

Fig. 3.10. Frequencies of hinged-fixed two bar frames —, $R = 200$; ······, $R = 100$; $- \cdot -$, $R = 50$

3.3 Vibrations of Three-Bar Frames

Also studied are vibrations of two types of plane frames made of three identical bars, for which $\alpha_1 = \alpha$, $\alpha_2 = \pi/2$, and $\alpha_3 = \pi - \alpha$. Both types have the top joints rigidly connected. One type has both ends hinged; the other has both ends fixed.

For both types, one may follow the same approach as for the two-bar frames to obtain the frequencies and normal functions. The eigenvalues k were computed and are presented in Figs. 3.11 and 3.12, respectively.

The principal modes may also be classified as symmetric and antisymmetric. The pattern of mode changes, however, is more complicated than those for the two-bar frames.

Figure 3.13 shows that both the second and third modes are symmetric for the frame with $\alpha = 60°$ and and $R = 100$, as indicated by the frequency curves in Fig. 3.11.

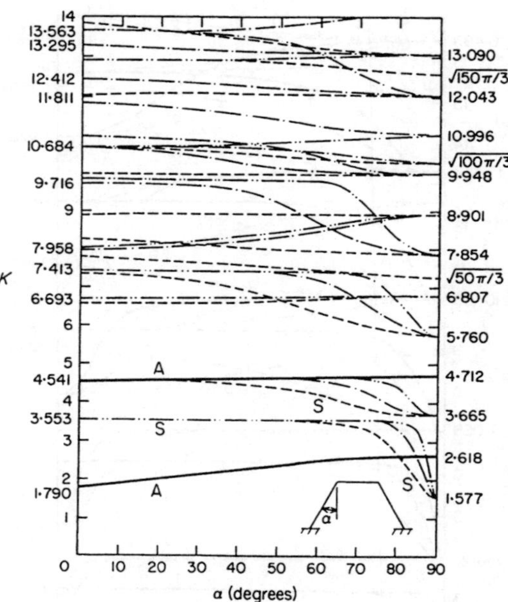

Fig. 3.11. Frequencies of three-bar frames with all joints fixed — ·· —, $R = 200$; — · —, $R = 100$; · · · · · ·, $R = 50$; —, for all R

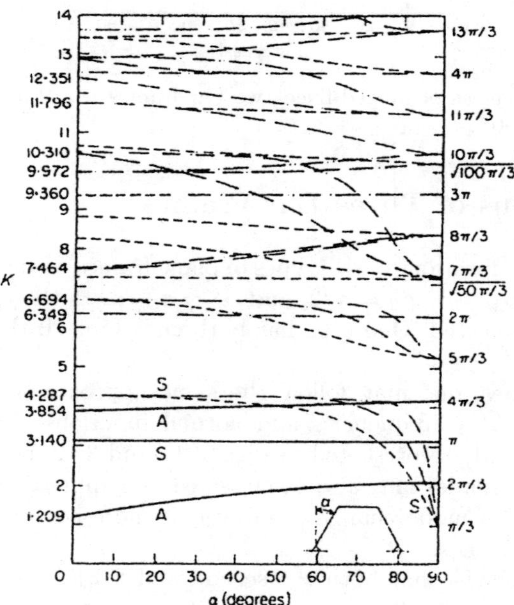

Fig. 3.12. Frequencies of bottom hinged, top fixed three-bar frames — ·· —, $R = 200$; — · —, $R = 100$; · · · · · ·, $R = 50$; —, for all R

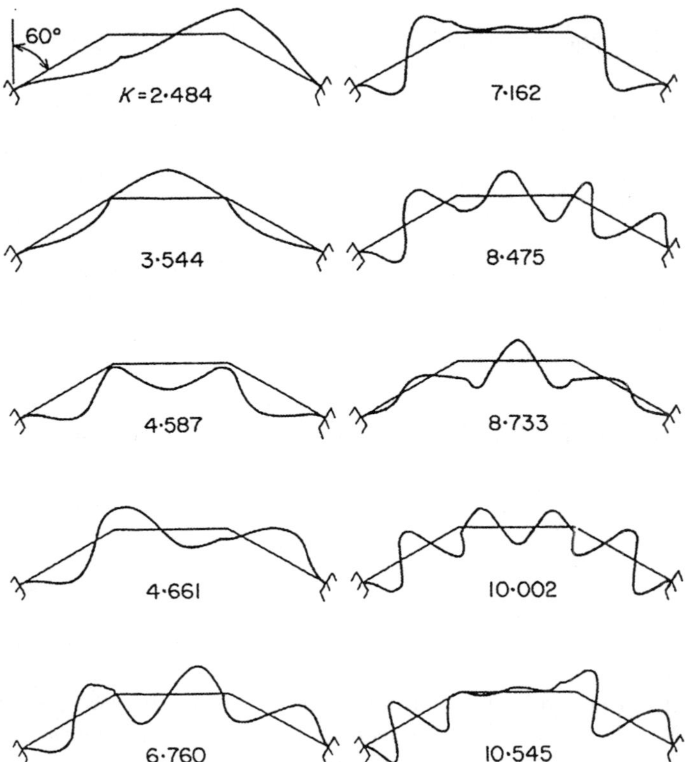

Fig. 3.13. Normal modes of a three-bar frames with all ends fixed $R = 100$; $\alpha = 60°$

For most range of α, the first mode is antisymmetric and independent to the slenderness ratio. The change of modes becomes unpredictable for frequencies higher than the first axial resonant frequency $\sqrt{\pi R/3}$.

3.4 Orthogonality of the Normal Functions

When dealing with transient and forced vibrations, the expansion of the normal functions may also be used if they form an orthogonal set. For this reason, the orthogonality of the normal functions is briefly presented.

The mth normal functions of the ith bar satisfy

$$(p_m/\beta_i)^2 W_{im} = W_{im}''''/R_i^2 \qquad (p_m/\beta_i)^2 U_{im} = -U_{im}'' \tag{3.45a,b}$$

By following the method used in the previous chapters, one may obtain the following equation:

$$
\begin{aligned}
(p_m^2 - p_n^2) \sum_i \frac{A_i E_i l_i}{\beta_i^2} \int_0^1 & (U_{im}U_{in} + U_{in}U_{im} + W_{im}W_{in} + W_{in}W_{im})\, d\eta \\
= \sum_i A_i E_i l_i & \left\{ \left[\left(\frac{W_{im}'''W_{in}}{R_i^2} - U_{im}'U_{in} \right) - \left(\frac{W_{in}'''W_{im}}{R_i^2} - U_{in}'U_{im} \right) \right] \right. \\
& \left. - \left[\frac{W_{im}''W_{in}' - W_{in}''W_{im}'}{R_i^2} \right] \right\} = 0 \qquad (3.46)
\end{aligned}
$$

Integration by parts has been performed on the right-hand side of the above equation. The terms in the brackets are vanished by applying proper boundary conditions. Thus one may conclude that the orthogonality of the normal functions takes the following form

$$
\sum_i \frac{A_i E_i l_i}{\beta_i^2} \int_0^1 (U_{im}U_{in} + W_{im}W_{in}) d\eta_i = 0 \qquad \text{for } m \neq n \qquad (3.47)
$$

Exercises

1. Check (3.35).
2. Derive the normal functions and check frequency (3.42) for a two-bar frame with all joints rigidly connected.
3. Derive the frequency equation for a portal frame with two ends fixed and the other two joints rigidly connected. All members are identical with $\alpha_1 = 0$, $\alpha_2 = \pi/2$ and $\alpha_3 = \pi$. Show the first four normal modes for this frame.

4

Vibrations of X-Braced Portal Frames

Because of devastation of structures due to earthquake and wind force, increasing attention has been paid to dynamic responses of braced structures [13, 22, 24, 45]. Extension of the formulation for vibrations of the two- and three-member frames of exact solutions in Chap. 3 to an x-braced portal frame is made in this chapter.

4.1 Free Vibration

The rectangular x-braced frame shown in Fig. 4.1 is considered. The two equations governing the transverse and axial vibrations for the ith (i =1, ..., 5) bar given in the last chapter are

$$\partial^4 \bar{W}_i / \partial x_i^4 + \left(\rho \bar{A}_i / EI_i\right) \partial^2 \bar{W}_i / \partial t^2 = 0 \tag{4.1a}$$

$$\partial^2 \bar{U}_i / \partial x_i^2 - (\rho / E) \, \partial^2 \bar{U}_i / \partial t^2 = 0 \tag{4.1b}$$

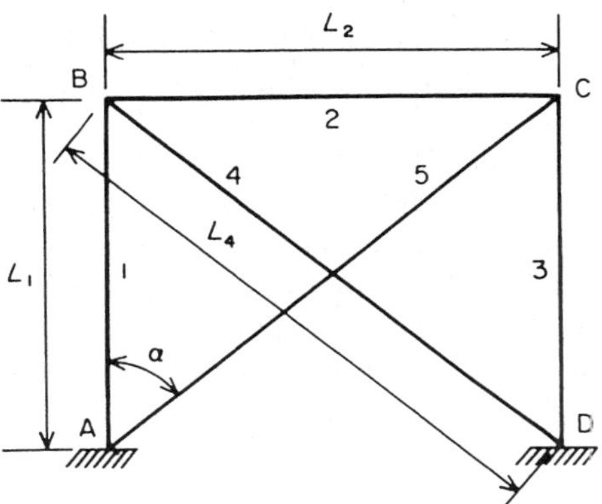

Fig. 4.1. An x-braced portal frame

where $\bar{W}_i(t, x_i)$ and $\bar{U}_i(t, x_i)$ are the transverse and axial displacement functions, respectively, of time t and the space coordinate x_i measured from one end of the ith bar along the centroid axis; $\bar{A}_i$ and I_i are cross-sectional area and area moment of inertia of the ith bar; and E and ρ are modulus of elasticity and mass density, respectively, which are considered to be the same for every bar.

Let

$$\bar{W}_i = W_i(\eta_i) L \cos pt \quad \text{and} \quad \bar{U}_i = U_i(\eta_i) L \cos pt \tag{4.2a,b}$$

where $L = L_1 = L_3$ is the length of the first and third bars, which are the columns; p is the radian frequency; and $\eta_i = x_i/L$. The solutions of (4.1) are

$$W_i = C_{i1} \sin KS_i\eta_i + C_{i2} \cos KS_i\eta_i + C_{i3} \sinh KS_i\eta_i + C_{i4} \cosh KS_i\eta_i \tag{4.3a}$$
$$U_i = C_{i5} \sin\left(K^2\eta_i/R\right) + C_{i6} \cos\left(K^2\eta_i/R\right) \tag{4.3b}$$

where

$$K^4 = \rho L^2 p^2 R^2/E \quad S_i^4 = r_1^2/r_i^2 \quad R = (L/r_1)^2 \tag{4.4a-c}$$

and $C_{is}(s = 1, 2, \ldots, 6)$ are constants to be determined by the end conditions; R is the slenderness ratio of the columns; and r_i is the radius of gyration of the ith bar.

The axial and shearing forces in each bar are, respectively,

$$\psi_i = \bar{A}_i E U_i' \quad \text{and} \quad \phi_i = -\left(\bar{A}_i E/R^2 S_i^4\right) W_i''' \tag{4.5a,b}$$

where a prime indicates the derivative with respect to n_i. The positive directions of ϕ_i and Ψ_i are shown in Fig. 3.1.

For free vibration, there are 30 homogeneous end conditions. At end A, $\eta_i = 0$, ($i = 1$ and 5), there are six boundary conditions:

$$W_1(0) = 0 \quad W_1'(0) = 0 \quad U_1(0) = 0 \quad W_5(0) = 0 \quad W_5'(0) = 0 \quad U_5(0) = 0 \tag{4.6a-f}$$

By using these six conditions, the numbers of constant C_{is} are reduced to 24. Let $C_{is} = C_j(j = 1, 2, \ldots, 24)$ which are to be determined by the following 24 conditions.

There are six similar conditions like (4.6a-f) at end D, where $\eta_3 = \eta_4 = 1$. At joints B and C, there are nine conditions at each joint: six conditions of continuity of horizontal, vertical displacements and slopes between each pair of the three bars meeting at the joints, and three equilibrium equations of horizontal and vertical forces and moments.

These 24 conditions may be expressed in a matrix form

$$[A_{ij}][C_j] = 0 \tag{4.7}$$

The elements of the coefficient matrix $[A_{ij}]$ $(i, j = 1, 2, \ldots, 24)$ are listed in the Appendix at the end of the book. For the non-trivial solution of C_j, the determinant of matrix $[A_{ij}]$ must vanish from which the frequency parameter $K(= T_1$ in the Appendix) may be determined.

In the search for frequencies, K, cautions must be taken. Particularly the first value, which is also the most important one, would be easily overlooked. For example, consider a frame with $\alpha = 30°$, $R = 50$, and $A_2 = A_4 = 1$, (A_i is normalized with A_1, such that A_i/A_1). If the increment $\Delta K = 0.1$ is taken, there is a dent near $K = 3.7$ as shown in Fig. 4.2. The horizontal x-axis is not crossed by the curve. When $\Delta K = 0.001$, the curve crosses the x-axis twice between 3.670 and 3.698 as shown in Fig. 4.3. Thus, there are two frequencies: 3.670 and 3.698. The mode curves for these two values are different.

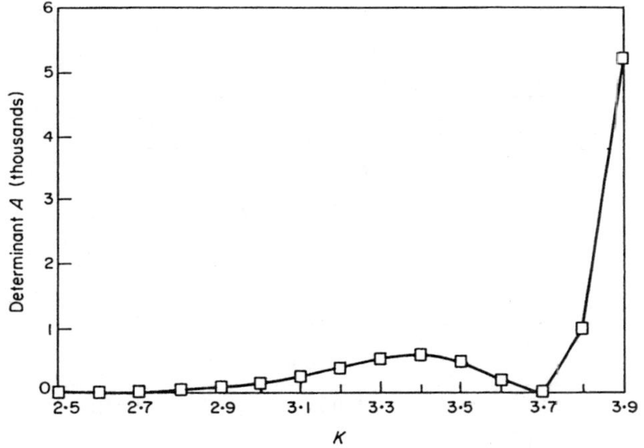

Fig. 4.2. Determinant A for $\alpha = 30°$, $R = 50$, $A_2 = A_4 = 1$

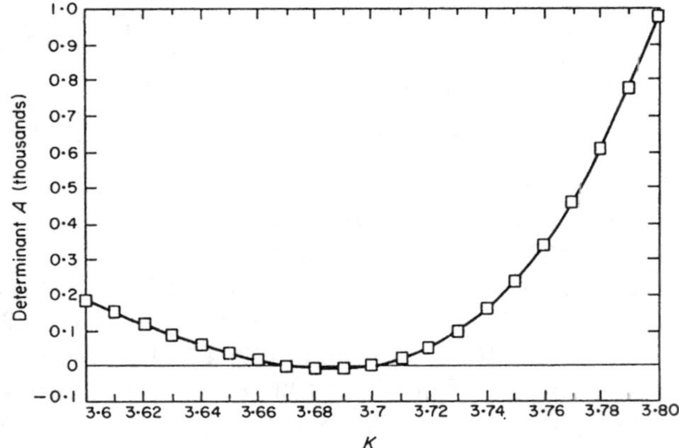

Fig. 4.3. Determinant A for $\alpha = 30°$, $R = 50$, $A_2 = A_4 = 1$

It was found that $\Delta K = 0.002$ was adequate and this was used in the computation of frequencies presented in this chapter.

The first five frequencies for different braced frames are obtained. They are listed in Tables 4.1 to 4.4 for the frequencies of K values for x-braced frames with different parameters: angles of the bracing bars, $\alpha = 15°$, $30°$, $45°$, $60°$ and $75°$, ($\alpha =$ B in the Appendix); slenderness ratios $R = 50$, 100 and 200; radius of gyration ratios $S_2 (S_i = r_1/r_i) = 0.5$ and 1.0; $S_4 = 1.0$, 2.0, and 4.0; and area ratios $A_2 = 1$ and 10. The frame, with $A_2 = 10$, represents a building frame with a heavy floor beam.

Table 4.1. First five frequencies of x-braced frames $A_2 = 1.0, S_2 = 1.0, A_4 = 1.0, S_4 = 1.0$

α				
15°	30°	45°	60°	75°
		$R = 50$		
3.268	3.670	3.022	2.129	1.112
4.146	3.698	3.135	2.240	1.177
4.635	4.407	3.983	2.541	1.241
4.659	4.504	4.269	3.590	1.892
5.020	5.138	4.701	3.761	1.972
		$R = 100$		
4.015	3.706	3.028	2.131	1.112
4.154	3.830	3.153	2.243	1.176
4.651	4.505	3.991	2.542	1.241
4.659	4.530	4.395	3.620	1.890
5.602	6.195	4.723	3.766	1.972
		$R = 200$		
4.153	3.708	3.029	2.131	1.113
4.283	3.861	3.157	2.242	1.176
4.660	4.505	3.993	2.542	1.241
4.663	4.555	4.414	3.623	1.890
6.740	6.259	4.728	3.768	1.972

The principal modes for the first five frequencies for $\alpha = 45°$, $R = 100$, and $A_2 = A_4 = S_2 = S_4 = 1.0$ are shown in Fig. 4.4.

For comparison, the frequencies of square portal frames without braces, $A_4 = A_5 = 0$, are obtained and listed in Tables 4.5 and 4.6. The first five modes of $\alpha = 45°$ and $R = 100$ are shown in Fig. 4.5. The frequencies of these modes may also be obtained from Fig. 3.11 for $\alpha = 0$ in the last chapter, except that the fourth frequency, 4.720, which is so close to the third one, is not listed there.

Table 4.2. First five frequencies of x-braced frames $A_2 = 10.0, S_2 = 0.5, A_4 = 1.0, S_4 = 1.0$

α				
15°	30°	45°	60°	75°
		$R = 50$		
2.627	3.089	3.170	2.355	1.220
4.563	4.082	3.332	2.360	1.223
4.603	4.142	3.436	3.048	1.683
4.724	4.722	4.595	3.295	2.029
4.761	4.760	4.744	3.925	2.036
		$R = 100$		
3.605	4.007	4.494	2.357	1.220
4.565	4.089	4.710	2.364	1.223
4.624	4.407	4.866	3.535	1.698
4.728	4.721	5.459	3.912	2.032
4.869	4.867	5.577	3.944	2.036
		$R = 200$		
4.455	4.086	3.337	2.358	1.220
4.565	4.089	3.341	2.363	1.223
4.672	4.708	4.710	3.604	1.702
4.728	4.721	4.716	3.920	2.032
5.487	6.243	5.521	3.946	2.036

Table 4.3. First five frequencies of x-braced frames $A_2 = 10.0, S_2 = 0.5, A_4 = 0.5, S_4 = 2.0$

α				
15°	30°	45°	60°	75°
		$R = 50$		
2.232	2.044	1.672	2.631	1.678
2.286	2.046	1.674	2.746	1.991
2.489	2.813	2.759	2.753	2.236
3.793	3.398	2.774	3.282	2.238
3.809	3.413	2.900	3.535	3.048
		$R = 100$		
2.279	2.048	2.776	3.517	1.694
2.286	2.050	2.778	3.531	2.234
3.348	3.392	3.874	3.553	2.236
3.793	3.402	3.887	3.705	2.795
3.827	3.936	4.048	4.320	3.331
		$R = 200$		
2.284	2.048	4.699	3.531	1.698
2.286	2.050	4.708	3.532	2.234
3.783	3.400	4.996	3.612	2.236
3.793	3.402	4.998	4.686	3.361
4.473	4.678	5.806	4.735	3.456

Table 4.4. First five frequencies of x-braced frames $A_2 = 10.0, S_2 = 0.5, A_4 = 0.1, S_4 = 4.0$

α				
15°	30°	45°	60°	75°
		$R = 50$		
1.142	1.700	2.056	1.840	1.392
1.144	1.702	4.427	3.269	1.676
1.884	2.080	4.725	3.338	2.338
1.898	4.718	4.754	3.340	2.340
1.958	4.762	5.276	4.516	2.444
		$R = 100$		
1.896	2.839	2.840	2.543	1.694
1.898	3.060	3.608	2.553	1.905
2.515	3.062	3.610	2.554	2.338
2.655	4.721	4.706	3.528	2.340
2.661	4.791	4.766	4.722	2.444
		$R = 200$		
3.397	3.740	3.966	3.572	1.697
3.412	3.742	4.708	3.607	2.336
3.457	3.941	4.798	4.722	2.338
4.728	4.721	5.274	4.728	2.412
4.881	4.842	5.276	4.730	2.414

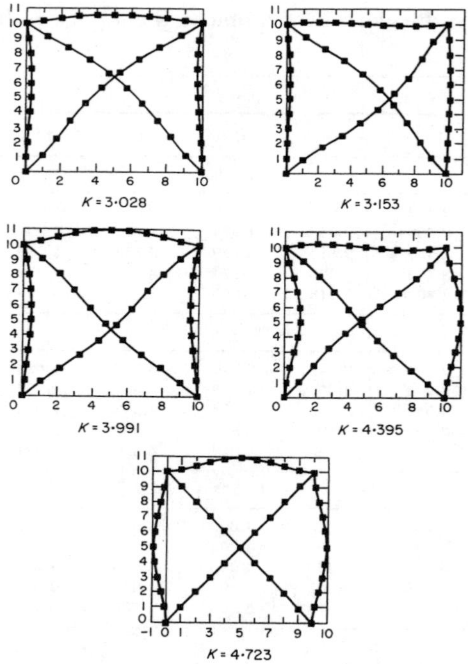

$K = 3{\cdot}028$ $K = 3{\cdot}153$

$K = 3{\cdot}991$ $K = 4{\cdot}395$

$K = 4{\cdot}723$

Fig. 4.4. First five modes of an x-braced frame with $\alpha = 45°, R = 100$ and $A_2 = S_2 = A_4 = S_4 = 1$

Table 4.5. First five frequencies of portal frames $A_2 = 1.0, S_2 = 1.0$

α				
15°	30°	45°	60°	75°
$R = 50$				
2.138	1.971	1.789	1.573	1.148
4.367	4.152	3.541	2.301	1.248
5.040	4.803	4.539	3.779	1.959
7.340	6.243	4.687	4.280	2.719
7.886	7.606	6.559	4.750	3.484
$R = 100$				
2.152	1.973	1.790	1.574	1.149
4.369	4.158	3.553	2.304	1.248
5.087	4.811	2.541	3.788	1.960
7.360	6.407	4.720	4.295	2.722
8.040	7.692	6.693	4.766	3.489
$R = 200$				
2.155	1.974	1.790	1.574	1.149
4.370	4.159	3.555	2.305	1.248
5.099	4.812	4.542	3.790	1.960
7.364	6.438	4.727	4.299	2.273
8.059	7.694	6.716	4.770	3.490

Table 4.6. First five frequencies of portal frames $A_2 = 10.0, S_2 = 0.5$

α				
15°	30°	45°	60°	75°
$R = 50$				
1.604	1.380	1.220	1.072	0.888
6.200	4.720	4.401	3.266	1.678
7.316	4.763	4.723	5.071	3.207
7.844	5.225	4.752	6.539	4.466
8.331	6.630	5.914	9.062	5.436
$R = 100$				
1.621	1.383	1.221	1.072	0.889
4.726	4.721	4.707	3.529	1.694
4.831	4.786	4.762	6.406	3.331
7.846	7.175	5.511	7.835	4.700
7.879	7.843	7.838	7.853	4.735
$R = 200$				
1.625	1.384	1.222	1.072	0.889
4.726	4.721	4.710	3.607	1.698
4.850	4.791	4.763	7.032	3.362
7.846	7.833	6.072	7.844	5.040
7.925	7.888	10.652	7.871	6.651

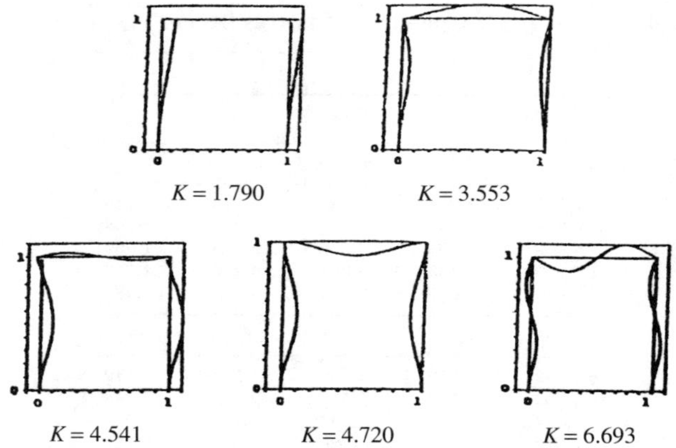

Fig. 4.5. First five modes of a portal frame with $A = 1,\ R = 100$

4.2 Forced Vibration

For forced vibration, the case of a frame acted upon by the horizontal force defined in the following equation is considered:

$$f = \bar{A}E\cos\left(\lambda^2 pt\right) \qquad 0 < \lambda < 1 \tag{4.8}$$

The force acts at the joint B in Fig. 4.1.

Let

$$\bar{w}_i = w_i\left(\eta_i\right)L\cos\left(\lambda^2 pt\right) \quad \text{and} \quad \bar{u}_i = u_i\left(\eta_i\right)L\cos\left(\lambda^2 pt\right) \tag{4.9a,b}$$

Then the solutions of (4.1) as given by (4.3) for free vibrations are still valid, by simply replacing the frequency parameter K by k, hence

$$k = \lambda K \tag{4.10}$$

where λ is a coefficient larger than zero and less than one (1). Thus

$$w_i = c_{i1}\sin kS_i\eta_i + c_{i2}\cos kS_i\eta_i + c_{i3}\sinh kS_i\eta_i + c_{i4}\cosh kS_i\eta_i \tag{4.11a}$$

$$u_i = c_{i5}\sin\left(k^2\eta_i/R\right) + c_{i6}\cos\left(k^2\eta_i/R\right) \tag{4.11b}$$

The frame has the same boundary and joint conditions except that the equation of equilibrium at joint B in the horizontal direction now becomes

$$\phi_1(1) - \psi_2(0) + \phi_4(0)\cos\alpha - \psi_4(0)\sin\alpha = \bar{A}_1 E \tag{4.12}$$

By using (4.5), the last equation may be reduced to

$$u_1'(1) + A_2\frac{1}{R^2S_2^4}w_2'''(0) + A_4u_4'(0)\cos\alpha + A_4\frac{1}{R^2S_4^4}w_6'''(0)\sin\alpha = 1 \tag{4.13}$$

Then (4.7) assumes the form

$$[A_{ij}]\,[c_j] = [b_i] \tag{4.14}$$

where $b_i = 0$ except $b_8 = 1$. For $k \neq K$, the inverse of the coefficient matrix $[A_{ij}]$ exists. Thus

$$[c_j] = [A_{ij}]^{-1}\,[b_j] \tag{4.15}$$

The solutions of (4.11) are now completely determined for a given value of λ.

The reactions to the horizontal force for frames with $\alpha = 45°$ and $R = 100$ are studied. Two sets of frames with different area ratios of the horizontal members are investigated for different bracing bars. The results shown in Figs. 4.6 and 4.7 are for the deflections, δ, (the horizontal displacement of joint $B \times 10^3/L$) versus external force frequency ratios, λ. Figures 4.6 and 4.7 are for $A_2 = 1.0$ and 10.0, respectively, showing the deflection δ versus λ of frames with different braces. Frames with $A_4 = 0$, 0.1, 0.5, and 1.0, each is labeled by A, B, C, and D, respectively, in both figures.

As a reference, the ratio of side deflection, δ, to the story height of a building according to the *Uniform Building Code* (UBC) [51] is limited to 0.5%, which is shown by a horizontal broken line, E, in both figures.

The configurations of the two x-braced frames of $A_2 = 1.0$ and 10.0 for different ratios of λ are shown in Figs. 4.8 and 4.9, respectively.

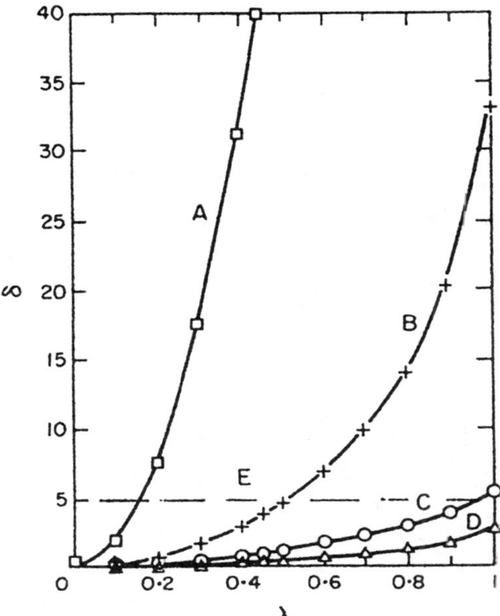

Fig. 4.6. Deflections vs. external force frequencies of braced frames with $A_2 = 1.0$, $S_2 = 1.0$; A : $A_4 = 0$; B: $A_4 = 0.1$, $S_4 = 4.0$; C: $A_4 = 0.5$, $S_4 = 2.0$; D: $A_4 = 1.0$, $S_4 = 1.0$; E: UBC limit

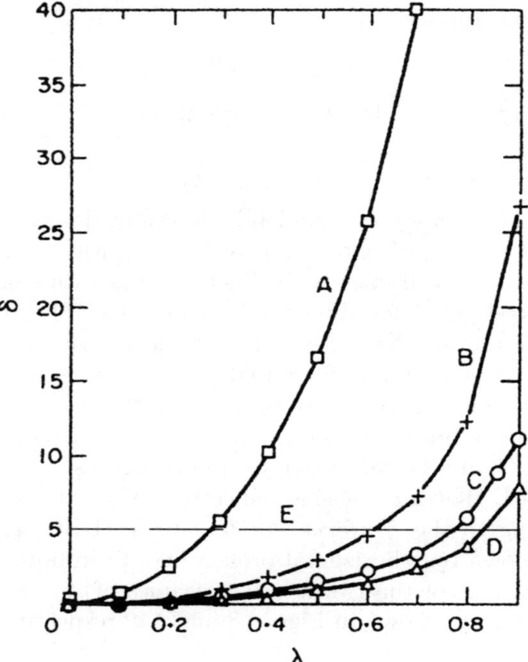

Fig. 4.7. Deflection vs. external force frequencies of braced frames with $A_2 = 10.0$, $S_2 = 0.5$, A: $A_4 = 0$; B: $A_4 = 0.1$, $S_4 = 4.0$; C: $A_4 = 0.5$, $S_4 = 2.0$; D: $A_4 = 1.0$, $S_4 = 1.0$; E: UBC limit

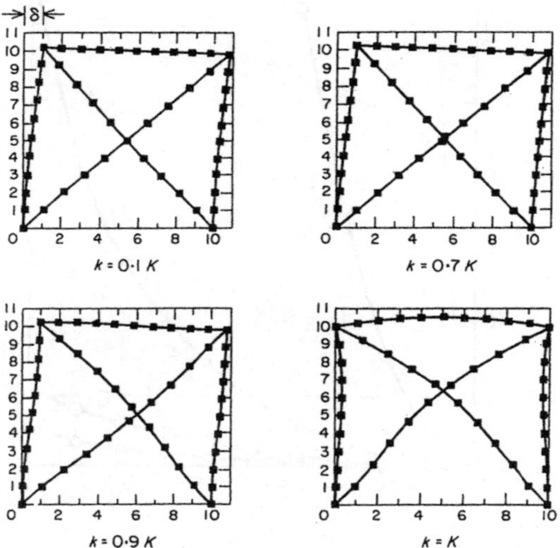

Fig. 4.8. Configurations of forced vibration of a braced frame of $\alpha = 45°$, $R = 100$, $A_2 = A_4 = S_2 = S_4 = 1, K = 3.028$

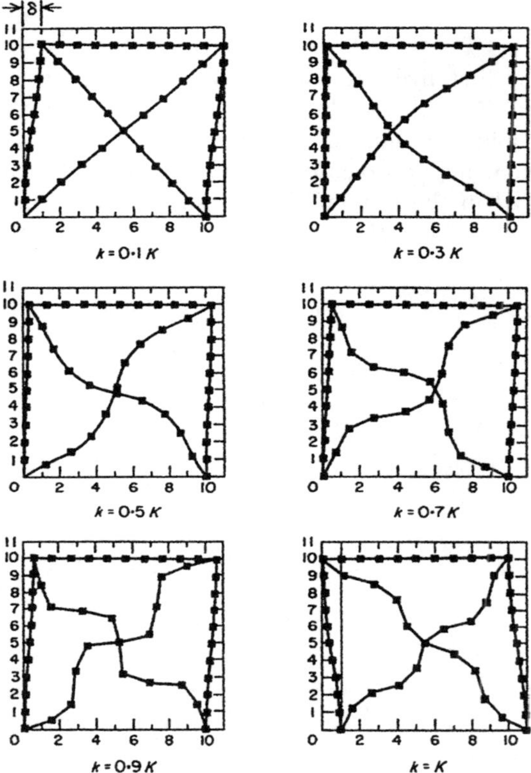

Fig. 4.9. Configurations of forced vibration of the braced frame of $\alpha = 45°$, $R = 100$, $A_2 = 10$, $S_2 = 0.5$, $A_4 = 0.1$, $S_4 = 4$, $K = 2.840$

When $\lambda = 1$, since K value has been rounded off, the inverse of the corresponding coefficient matrix still exists.

For $A = 10$, the floor beam remains straight for all k values as shown in Fig. 4.9, and the four deformations for the braces are quite severe after $\lambda > 0.3$. This may explain that the curves C and D in Fig. 4.6 are lower than the corresponding curves in Fig. 4.7.

There are a number of computer programs for vibration analysis of frames available [18], but none of them includes the effect of axial vibrations, which could not be neglected for frames with diagonal members. The results presented in this study may be considered as exact in the framework of small vibration.

Exercises

1. Determine the first five frequencies and normal modes for the frame of Fig. 4.1, $\alpha = 45°$. All members are identical except that member 5 is removed. All joints are hinged.
2. Determine the configurations for $\lambda = 0.1$, 0.3, 0.5, 0.7, and 0.9 when the frame, specified in the last problem, is subjected to the horizontal force defined by (4.8) at joint B.

5

Vibrations of X-Braced Multi-Story Frames

The study of vibrations for the x-braced portal frames in the last chapter is extended to high-rise x-braced multi-story frames in this chapter. Frequencies for two- to five-story frames are given. Forced vibrations for three-story frames are analyzed. Investigations are also made for the effects of horizontal ground motion for three- to five-story frames. Comparison of the present approach with available frequencies of a three-story rectangular frame is made.

5.1 Formulation

Consider the N-story, x-braced elastic frame shown in Fig. 5.1. The configuration of the frame may be defined by the height of the column, L_n $(n = 1, 2, \ldots,$ N), of each story and the angle, α, between the braces and columns of the first story as shown in Fig. 5.1.

The two equations governing the transverse and axial vibrations for the ith (i = 1, 2, ..., 5N) bar elements as derived in Chap. 2 are

$$\frac{\partial^4 \bar{W}_i}{\partial x_i^4} + \frac{\rho_i \bar{A}_i}{E_i I_i} \frac{\partial^2 \bar{W}_i}{\partial t^2} = 0 \tag{5.1a}$$

$$\frac{\partial^2 \bar{U}_i}{\partial x_i^2} - \frac{\rho_i}{E_i} \frac{\partial^2 \bar{U}_i}{\partial t^2} = 0 \tag{5.1b}$$

in which ρ_i, E_i, $\bar{A}_i$, L_i and I_i are mass density, modulus of elasticity, cross-sectional area, length and area moment of inertia of the ith bar, respectively. Let

$$\bar{W}_i = W_i(\eta_i) L_1 \cos pt \tag{5.2a}$$

$$\bar{U}_i = U_i(\eta_i) L_1 \cos pt \tag{5.2b}$$

in which p = circular frequency and $\eta_i = x_i/L_1$. Then the solutions of (5.1) are

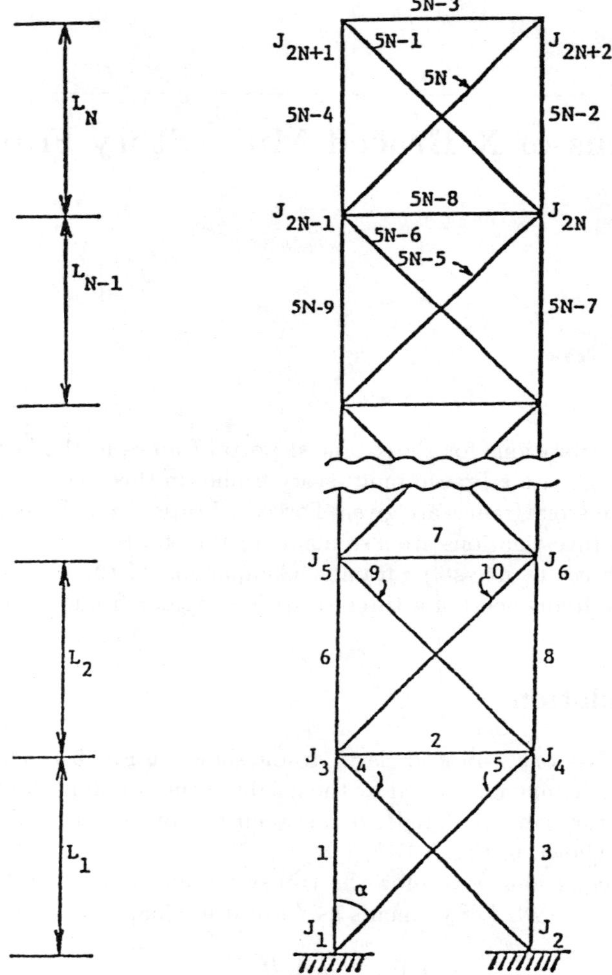

Fig. 5.1. An N-story x-braced frame

$$W_i = C_{i1} \sin(Y_i) + C_{i2} \cos(Y_i) + C_{i3} \sinh(Y_i) + C_{i4} \cosh(Y_i) \quad (5.3a)$$

$$U_i = C_{i5} \sin(Z_i) + C_{i6} \cos(Z_i) \quad (5.3b)$$

in which

$$Y_i = KM_iS_i\eta_i \quad \text{and} \quad Z_i = K^2M_i^2\eta_i/R \quad (5.4a,b)$$

$$K = \left(\frac{\rho_1 L_1^2 p^2 R^2}{E_1}\right)^{1/4} = \text{frame frequency} \quad (5.5a)$$

$$S_i = \left(\frac{r_1}{r_i}\right)^{1/2} = \text{stiffness factor} \quad (5.5b)$$

$$M_i = \left(\frac{E_1 \rho_i}{E_i \rho_1}\right)^{1/4} = \text{material constant} \tag{5.5c}$$

$$R = \frac{L_1}{r_1} = \text{slenderness ratio of column 1} \tag{5.5d}$$

$$r_i = \text{radius of gyration} \tag{5.5e}$$

$$A_i = \frac{\bar{A}_i}{A_1} = \text{area ratio} \tag{5.6}$$

and C_{is} $(s = 1, 2, \ldots, 6)$ are constants to be determined by the end conditions.

5.2 Free Vibration

Consider the frame shown in Fig. 5.1. It is rigidly joined and fixed at end joints, J_1 and J_2. At each end joint, there are six homogeneous geometrical conditions of W_1, U_1, W_5, U_5, W_1', and W_5'. At each of the intermediate joints, J_3, J_5, ..., $J_{2\mathrm{N}-1}$, there are twelve conditions of compatibility and three equations of equilibrium. At the top two joints, $J_{2\mathrm{N}+1}$ and $J_{2\mathrm{N}+2}$, there are six conditions of compatibility and three equations of equilibrium. Details of these compatibility and equilibrium equations are given in Chaps. 3 and 4. Thus, for a N-story frame, there are 5N bars and each bar has six constants, C_{is}. In all, there are 30N constants to be determined by $2[6 + (\mathrm{N} - 1) \times 15 + 9] = 30\mathrm{N}$ equations.

Let $C_{is} = C_j (j = 1, 2, \ldots, 30\mathrm{N})$, the 30N equations may be expressed in a matrix form

$$[A_{mj}]\,[C_j] = [b_j] \quad (m = 1, 2, \ldots 30\mathrm{N}) \tag{5.7}$$

For free vibration, (5.7) is homogeneous. In order to have non-trivial solutions of vector C_j, the determinant of the coefficient matrix must vanish. This yields the eigenvalues or frame frequencies, K. For each value of K, the mode of free vibration may be determined.

In the process to seek the eigenvalues, K, cautions must be taken to ensure the increment value of ΔK to be small enough so that no eigenvalues would be skipped as it was explained in Chap. 4. In the numerical examples given in this study, the increment of K, $\Delta K = 0.002$ was used.

5.3 Vibrations due to Horizontal Forces and Ground Motion

For the steady forced vibration, consider the case when the frame is acted upon by a horizontal compressive force of (5.8) at the nth floor joint

$$F_n = f_n \bar{A}_i E_i \cos\left(\lambda^2 pt\right) \quad 0 < \lambda < 1 \tag{5.8}$$

in which f_n is a dimensionless force. Since the forces are acting at the joints only, (5.1) are still valid. Let the solutions of these equations be

$$\bar{w}_i = w_i(\eta_i) L_1 \cos(\lambda^2 pt) \tag{5.9a}$$
$$\bar{u}_i = u_i(\eta_i) L_1 \cos(\lambda^2 pt) \tag{5.9b}$$

Then the solutions given by (5.3) for free vibrations are valid, simply by replacing the frame frequency K by k, so that

$$k = \lambda \tag{5.10}$$

where $\lambda (0 \leq \lambda < 1)$ is a coefficient. Hence,

$$w_i = c_{i1} \sin y_i + c_{i2} \cos y_i + c_{i3} \sinh y_i + c_{i4} \cosh y_i \tag{5.11a}$$
$$u_i = c_{i5} \sin z_i + c_{i6} \cos z_i \tag{5.11b}$$

in which

$$y_i = k M_i S_i \eta_i \tag{5.12a}$$
$$z_i = k^2 M_i^2 \eta_i / R \tag{5.12b}$$

Now (5.7) becomes

$$[A_{mj}][C_j] = [b_j] \tag{5.13}$$

in which

$$b_j = f_n = \text{ force coefficient} \tag{5.14}$$

and appears in the horizontal equilibrium equations at joints where the forces are applied such as J_3, J_5, ..., J_{2n+1}; otherwise, $b_j = 0$.

For a given value of $\lambda (\lambda < 1)$, the inverse of the coefficient matrix $[A_{mj}]$ exists. Then

$$[C_j] = [A_{mj}]^{-1} [b_j] \tag{5.15}$$

The solutions of (5.11) are now completely determined for a given value of λ.

For the steady vibrations of the frame due to horizontal sinusoidal ground vibrations, a similar approach can be employed. In this case, the conditions at all joints remain the same as in the free vibrations except that the four boundary conditions at the bottom ends, J_1 and J_2, now become

$$\bar{w}_1(0) = G L_1 \cos(\lambda^2 pt) \tag{5.16a}$$
$$\bar{w}_5(0) \cos\alpha + \bar{u}_5(0) \sin\alpha = G L_1 \cos(\lambda^2 pt) \tag{5.16b}$$
$$\bar{w}_3(1) = G L_1 \cos(\lambda^2 pt) \tag{5.16c}$$
$$-\bar{w}_4\left(\frac{L_1}{\cos\alpha}\right) \cos\alpha + \bar{u}_4\left(\frac{L_1}{\cos\alpha}\right) \sin\alpha = G L_1 \cos(\lambda^2 pt) \tag{5.16d}$$

in which the dimensionless amplitude, G, is restricted to < 1.

Replacing the corresponding b_j in (5.14) by these four conditions to solve for $C_j = c_{is}$, the solutions of (5.11) are completed for the given values of G and λ.

5.4 Numerical Examples

In order to investigate the characteristics of high-rise x-braced frames, five sets of numerical examples are presented in the following: (1) natural frequencies of 2- to 5-story frames, (frequencies of 1-story frame are available in Chap. 4); (2) principal modes of free vibrations of two 3-story frames; (3) reactions of forced vibrations of a 3-story frame; (4) responses of 3-, 4-, and 5-story frames due to ground vibrations; and (5) comparison of the present approach with a discrete system method for a 3-story rectangular unbraced frame. For simplicity, all the frames are made of one material, $M_i = 1$.

5.4.1 Natural Frequencies of 2- to 5-Story Frames

The first three frequencies of the free vibrations of 2- to 5-story x-braced frames are presented in Tables 5.1 to 5.4, respectively, for $A_i = S_i = 1$.

Table 5.1. First three frequencies of 2-story x-braced frames for all $A_i = 1,\ S_i = 1$

$\alpha = 15°$ (1)	$\alpha = 30°$ (2)	$\alpha = 45°$ (3)	$\alpha = 60°$ (4)	$\alpha = 75°$ (5)
		$(a)\ R = 50$		
1.882	2.494	2.713	2.082	1.108
3.789	3.571	2.915	2.095	1.120
4.026	3.755	3.131	2.239	1.149
		$(b)\ R = 100$		
2.591	3.225	2.909	2.090	1.110
4.030	3.586	2.932	2.103	1.121
4.170	3.827	3.182	2.240	1.149
		$(c)\ R = 200$		
3.418	3.520	2.932	2.092	1.110
4.031	3.589	2.936	2.105	1.121
1.171	3.828	3.183	2.240	1.149

The first frequency of a frame is also the most important one. The first frequencies of the frames of 1- to 5-story for $A_i = S_i = 1$ and $\alpha = 10°$ increased by every 5° up to 80° were computed. The results are presented in Figs. 5.2, 5.3, and 5.4 for $R = 50$, 100, and 200, respectively.

These figures show that the frame frequency, K, gets lower as the number of stories, N, increases. This indicates that as the frame gets higher, it becomes more flexible, as it was predicted by an approximate method in [50] which

Table 5.2. First three frequencies of 3-story x-braced frames for all $A_i = 1, S_i = 1$

$\alpha = 15°$ (1)	$\alpha = 30°$ (2)	$\alpha = 45°$ (3)	$\alpha = 60°$ (4)	$\alpha = 75°$ (5)
		(*a*) $R = 50$		
1.298	1.781	2.116	2.046	1.101
2.863	3.352	2.865	2.063	1.107
3.996	3.514	2.877	2.148	1.131
		(*b*) $R = 100$		
1.814	2.473	2.756	2.058	1.102
3.647	3.501	2.907	2.081	1.109
4.009	3.554	2.938	2.150	1.132
		(*c*) $R = 200$		
2.525	3.233	2.859	2.060	1.102
3.955	3.551	2.916	2.086	1.110
4.010	3.561	3.025	2.150	1.132

Table 5.3. First three frequencies of 4-story x-braced frames for all $A_i = 1,\ S_i = 1$

$\alpha = 15°$ (1)	$\alpha = 30°$ (2)	$\alpha = 45°$ (3)	$\alpha = 60°$ (4)	$\alpha = 75°$ (5)
		(*a*) $R = 50$		
0.984	1.368	1.664	1.855	1.093
2.289	2.956	2.806	2.042	1.106
3.390	3.434	2.836	2.050	1.126
		(*b*) $R = 100$		
1.381	1.923	2.310	2.037	1.093
3.097	3.444	2.848	2.075	1.109
3.867	3.536	2.891	2.120	1.127
		(*c*) $R = 200$		
1.943	2.660	2.813	2.040	1.093
3.835	3.463	2.907	2.083	1.110
3.958	3.549	2.951	2.121	1.127

Table 5.4. First three frequencies of 5-story x-braced frames for all $A_1 = 1,\ S_1 = 1$

$\alpha = 15°$ (1)	$\alpha = 30°$ (2)	$\alpha = 45°$ (3)	$\alpha = 60°$ (4)	$\alpha = 75°$ (5)
		$(a)\ R = 50$		
0.789	1.106	1.361	1.569	1.089
1.891	2.525	2.724	2.016	1.105
2.921	3.289	2.771	2.031	1.122
		$(b)\ R = 100$		
1.111	1.561	1.913	2.021	1.089
2.609	3.307	2.822	2.069	1.109
3.696	3.453	2.874	2.088	1.124
		$(c)\ R = 200$		
1.569	2.190	2.605	2.029	1.089
3.469	3.429	2.831	2.081	1.110
3.925	3.542	2.901	2.108	1.125

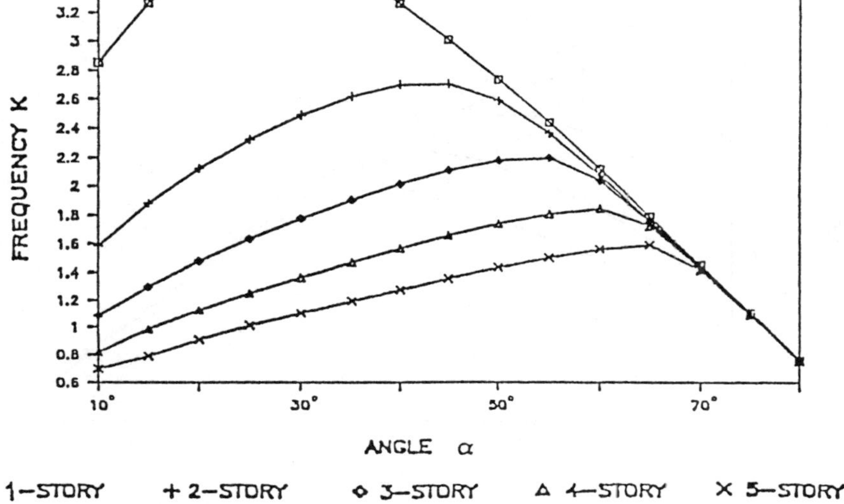

Fig. 5.2. First frequencies of 1- to 5-story braced frames $R = 50$

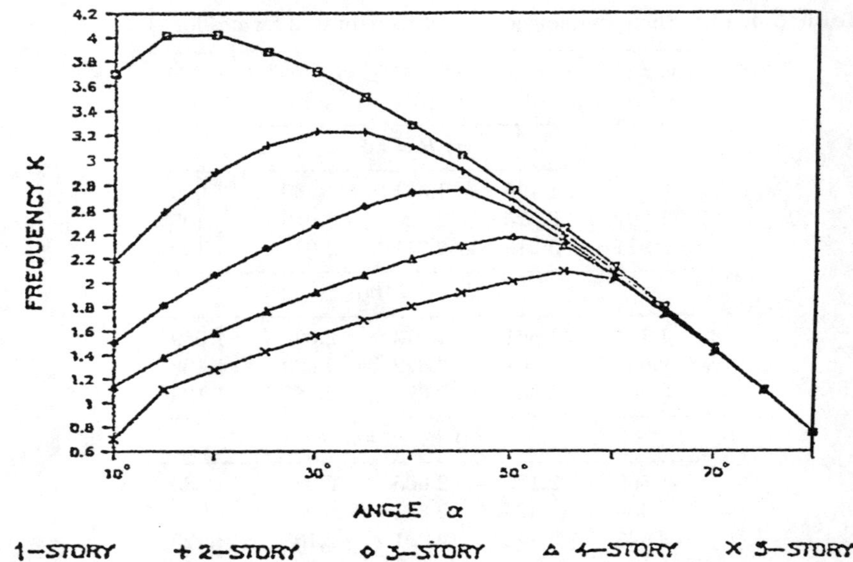

Fig. 5.3. First frequencies of 1- to 5-story braced frames $R = 100$

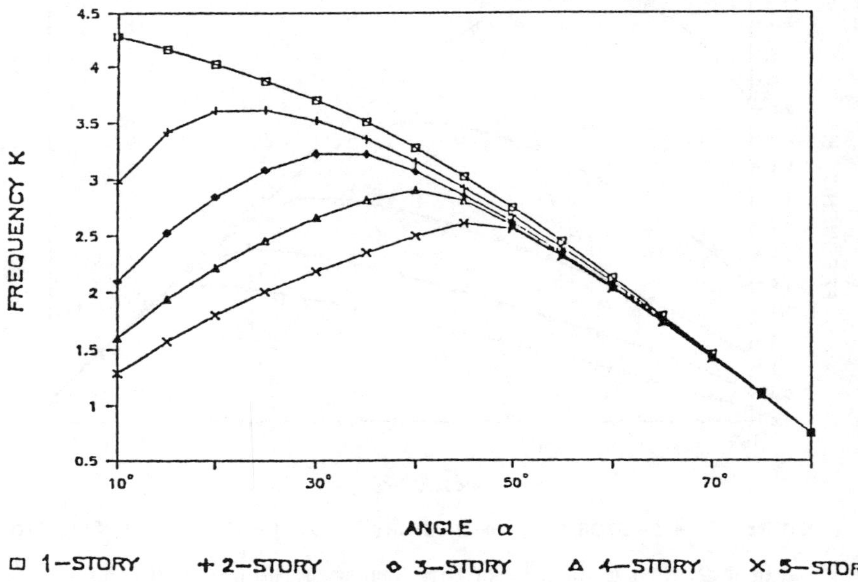

Fig. 5.4. First frequencies of 1- to 5-story braced frames $R = 200$

indicates that building frequency is inversely proportional to N or N^2. For a given frame, there is an optimal or a maximum value of K due to different α.

For frames with a small angle, ($\alpha < 25°$ for $R = 50$), the K value drops. This is due to the increase of the flexibility of the frames as a whole. Then the frame may vibrate as a cantilever beam.

On the other hand, for the frames with a large angle, ($\alpha > 65°$), K value is controlled by the vibrations of the horizontal and diagonal members which become very flexible due to the increase of the span length. Thus, the K values are independent to the number of stories. This may be seen by the following fact. For instance, for a frame of $\alpha = 75°$ and $R = 50$, the K value obtained from Fig. 5.2 is proximately 1.1. The frequency of the horizontal beam is $1.1 \times \tan 75° = 4.105$ which lies between the frequencies of a single beam $K = 4.730$ with fixed ends, and $K = 3.142$ with hinged ends (seen from Figs. 2.3 and 2.8, $\alpha = 0$, in Chap. 2).

5.4.2 Modes of Free Vibrations of Two 3-Story Frames

Presented in Fig. 5.5 is a set of the first five modes of free vibrations of a 3-story frame of $R = 100$ and $\alpha = 45°$. It shows that only the fourth mode is symmetric; the rest are asymmetric. There is no cantilever-beam type of mode exhibited. This is quite different from such frames without braces. As it was shown in [50], the first three asymmetric modes of a 3-story unbraced frame were cantilever-beam type and the fourth mode was symmetric. For different area ratios, the mode pattern changes.

In large scale lattice structures, equivalent continuum models were used [2,30]. The structure would behave as a cantilever beam. To see whether such a cantilever-beam type of mode ever existed in x-braced frames at the basic frequencies, another set of the first three modes of a frame with $R = 100$ and

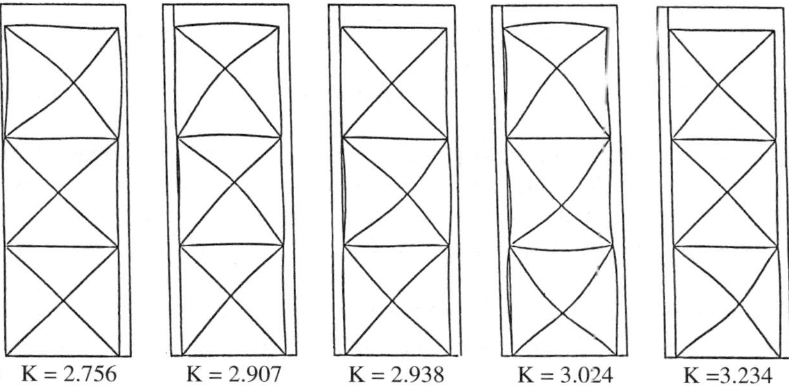

Fig. 5.5. First five modes of a 3-story braced frame $\alpha = 45°$, $R = 100$ and $A_i = S_i = 1$

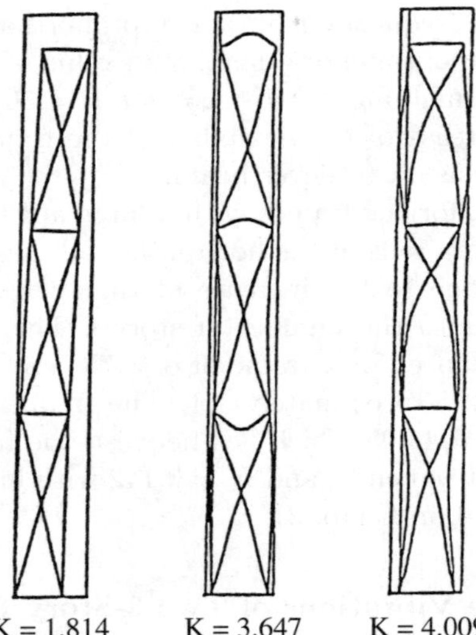

Fig. 5.6. First 3 modes of a 3-story braced frame $\alpha = 15°$, $R = 100$ and $A_i = S_i = 1$

$\alpha = 15°$ is presented in Fig. 5.6 which indicates that only the first mode is cantilever-beam type and the second and third modes are symmetric.

5.4.3 Forced Vibrations of a 3-Story Frame

The steady state of forced vibrations of a 3-story frame with $A_i = S_i = 1$, $R = 100$, and $\alpha = 45°$ subjected to horizontal compressive forces $f_n = 1$ (n = 3, 5, 7) at joints J_3, J_5, and J_7 is used as an illustrative example. The normalized deflected configurations are presented in Fig. 5.7 when the frequency parameter of the external force, $\lambda = 0.5$ and 0.9, and for $K = 2.756$, the first frequency. The deflection is normalized with the static deflection of the first floor joint, J_3, which was obtained by assuming a very small value of λ (= 0.001). Figure 5.7 indicates that the most destruction occurs at the third floor at $\lambda = 0.9$.

The spectra of frequency ratio, λ, of the external forces versus the floor joint in horizontal deflections of the three floors are depicted in Fig. 5.8, which indicates that for $\lambda < 0.3$, there is not much difference between the static and dynamic deflections. However, for $\lambda > 0.4$, the dynamic response increases rather rapidly.

Presented in Fig. 5.8 is also the transverse amplitude at the middle of the 14th bracing bar. For small λ, this amplitude is almost equal to the deflection of joint 5 for $\lambda < 0.4$; as λ approaches the resonant value, it increases at a rate faster than the joint displacements, hence these bracing bars become the most critical elements of the frame.

Fig. 5.7. Deflected 3-story braced frames due to side forces

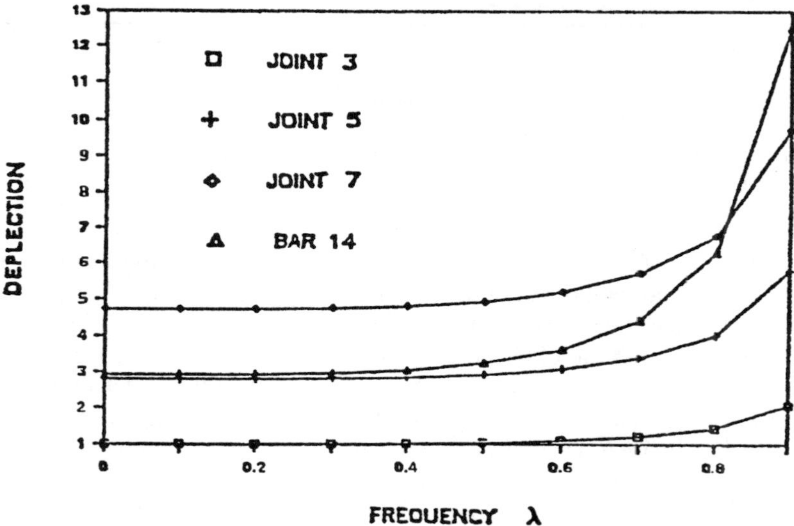

Fig. 5.8. Joint displacements vs. frequency ratio λ of forced vibration for a 3-story braced frame

5.4.4 Response to Horizontal Ground Vibrations for 3-, 4- and 5-Story Frames

An investigation is made for the steady vibrations of 3-, 4-, and 5-story x-braced frames with $A_i = S_i = 1$, $\alpha = 45°$ and $R = 100$ subjected to harmonic ground vibrations as defined in (5.16a-d). The amplitude of ground motion, $G = 0.1$, is used.

The spectra of λ versus the horizontal deflection of the third, fifth, and seventh floor joints of a 3-story frame is presented in Fig. 5.9, which shows

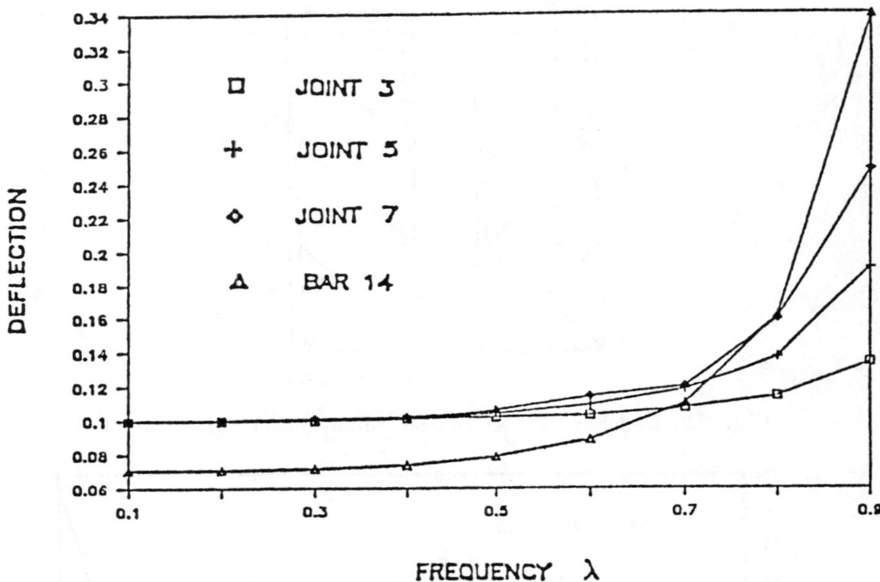

Fig. 5.9. Joint displacements vs. frequency ratio of horizontal ground vibration for 3-story x-braced frame

that with $\lambda < 0.4$, the deflections of the three joints are almost equal. This indicates that the frame responds as a rigid body. However, when $\lambda > 0.4$, the dynamic effect becomes significant. Again, as in the forced vibrations, the amplitude at the middle of the 14th bracing bar increases more rapidly than the deflections of the floor joints, as λ approaches the resonant value.

The configurations of the 3-, 4-, and 5-story frames are given in Fig. 5.10 at $k = 0.9K$, where K is the first frequency of free vibrations of each frame.

All three configurations show that the most destruction occurs at the top story. The relative destruction decreases as the number of stories increases. This may be due to the fact that as the frame gets higher, it is more flexible and more energy would be absorbed by the other stories of the frame.

5.4.5 Comparison with a Rectangular Unbraced Frame

The 3-story unbraced rectangular frame studied by Vertes in [53] by using a discrete method is taken here for comparison with the present approach.

The frame has $\bar{A}_i = 0.3\,\mathrm{m}^2$, $I_i = 0.01\,\mathrm{m}^4$, $L_n = 6\,\mathrm{m}$ $(n = 1,2,3)$, $E_i = 2 \times 10^7\,\mathrm{kN/m}^2$, $\rho_i = 2.584\,\mathrm{kg/m}^3$ $(i = 1$ to 9, the number of members), and $\alpha = 53.13°$. For simplicity, rectangular cross-sections are assumed, hence $R = 32.86$. The first four frequencies given in [53] are compared with the results obtained from the present approach as listed in Table 5.5.

Fig. 5.10. Deflected frames near resonance of horizontal ground vibration

Table 5.5. First four frequencies of a 3-story frame $A_i = 1,\ S_i = 1,\ R = 32.86$ and $\alpha = 53.13°$

Modes	Vertes (K)	Present (K)
1	0.907	0.890
2	1.665	1.657
3	2.270	2.263
4	2.928	2.705

The frequencies obtained by the present method are slightly lower as expected. The discrepancy between the fourth frequency figures is relatively large. This indicates that in the computation of frequencies by discrete system method, the error will get bigger as the modes are getting higher.

This fact may be seen from the comparison of the frequencies of a cantilever beam obtained by the finite element method with those from the continuum mode in [15].

The four mode configurations presented in [53], however, are almost identical with those obtained by the present approach, hence they are not shown here.

5.5 Discussion

The present study has shown that the classic elastic continuum approach to solve the frame as well as truss vibration problems is feasible. In comparison to the discrete methods, such as the finite element method (FEM), the present approach has the following advantages:

1. The solutions of transverse and axial vibrations are well-known. Thus, the present method is straightforward. It would be quite involved to form the stiffness matrix of x-braced frames by using the FEM.
2. In the present approach, both the transverse and axial vibrations are included, and each bar has six constants of integration for the basic as well as for the higher modes. Therefore, the formulation of the computer programs can be written once for all. In FEM, there are also six unknowns of end displacements and forces including the bending and axial deformations for each element of the basic mode. For the second and higher modes, each bar must be broken into two, three or more elements. Hence the numbers of unknowns will be proportionally increased. Thus a computer program has been written only for up to certain degrees of frequencies. For any higher modes, the program has to be modified accordingly.
3. After all, the present method results in exact solutions in the framework of small vibration theory, and FEM yields only approximate values. The accuracy of FEM depends on the number of elements chosen.

It may be concluded that for the analysis of small in-plane vibrations of plane frames, as well as plane trusses, the present continuum model yields simple formulations and accurate results. The x-braces of rectangular frames are known to be very effective in resistance of lateral forces. The present study provides a quantitative analysis. Since the results presented are based on exact solutions, they may be used as the basis of comparison for discrete numerical methods.

Some conclusions may also be made from observations of the analytic results obtained:

1. The basic cantilever-beam type of vibration mode of unbraced rectangular building frames may not take place for x-braced frames if the angle of α is not too small.
2. In general, the basic frequency gets lower as the number of stories increases. However, for frames with a large angle of α, the frequency is independent to the number of stories.
3. In the response of horizontal ground vibrations, the most destruction occurs on the top floor.
4. In the near resonant frequency of forced vibrations as well as due to horizontal ground motions, the diagonal braces are the most critical members of the frame, particularly in the top story.

Exercises

1. Determine the first three frequencies and normal modes of the two-story frame made of the frame of problem 1 in Chap. 4. (Members 5 and 9 are removed.)
2. Determine the deflections of the first and second floor joints for the two-story frame in the last problem due to ground vibrations, $G = 0.1$. Present the results as given in Fig. 5.9.

6

Effect of Rotatory Inertia and Shear Deformation on Vibration of Inclined Bars with End Constraint

The effect of rotatory inertia and shear deformation on the vibration of the inclined bar with the end constraint studied in Chap. 2 is examined here. The basic equations are derived by the variation method. The present work may be considered as an extension of the investigation of that chapter; therefore, the same nomenclature are used herein with some specified additions.

6.1 Formulation

Let ϕ be the angle of rotation due to bending only. Then the kinetic energy and potential energy of an inclined bar undergoing free vibration may be given as, respectively,

$$T = \frac{1}{2}\frac{AE}{\beta^2}\int_{t_0}^{t_1}\int_0^1 \left[\dot{U}^2 + \dot{W}^2 + \dot{\phi}^2/R^2\right] d\eta dt \tag{6.1a}$$

$$V = \frac{1}{2}AE\int_{t_0}^{t_1}\int_0^1 \left[\left(U,_\eta + 1/2W,_\eta^2\right)^2 + \frac{1}{R^2}\phi,_\eta^2 + s^2\left(W,_\eta - \phi\right)^2\right] d\eta dt \tag{6.1b}$$

where

$$s^2 = \mathrm{K}'G/E \tag{6.2}$$

in which K' is Timoshenko's shear coefficient [14, 26, 29, 38, 47].

The upper end of the bar is moveable along a straight path, which requires

$$U\sin\alpha + W\cos\alpha = 0 \quad \text{at} \quad \eta = 1 \tag{6.3}$$

By combining expressions (6.1 and 6.3), a functional equivalent to Lagrangian function may be formed:

$$L = T - V + \lambda\int_{t_0}^{t_1}\left[U\sin\alpha + W\cos\alpha\right]_{\eta=1} dt \tag{6.4}$$

Applying the following three Lagrange-Euler equations

$$\left(\frac{\partial L}{\partial \dot{U}}\right)^{\cdot}+\left(\frac{\partial L}{\partial U_{,\eta}}\right)_{,\eta}=0 \quad \left(\frac{\partial L}{\partial \dot{W}}\right)^{\cdot}+\left(\frac{\partial L}{\partial W_{,\eta}}\right)_{,\eta}=0 \quad \left(\frac{\partial L}{\partial \dot{\phi}}\right)^{\cdot}-\frac{\partial L}{\partial \phi}+\left(\frac{\partial L}{\partial \phi_{,\eta}}\right)_{,\eta}=0 \tag{6.5a-c}$$

to functional (6.4) results in three equations of motion:

$$\left(1/\beta^2\right)\ddot{U}-\left(U_{,\eta}+1/2W_{,\eta}^2\right)_{,\eta}=0 \tag{6.6a}$$

$$\left(1/\beta^2\right)\ddot{W}-\left[\left(U_{,\eta}+1/2W_{,\eta}^2\right)W_{,\eta}\right]_{,\eta}-s^2\left(W_{,\eta}-\phi\right)_{,\eta}=0 \tag{6.6b}$$

$$\left(1/R^2\beta^2\right)\ddot{\phi}-\left(1/R^2\right)\phi_{,\eta\eta}-s^2\left(W_{,\eta}-\phi\right)=0 \tag{6.6c}$$

Through the variation method, the following boundary conditions are obtained:

$$U=0 \qquad W=0 \quad \text{at} \quad \eta=0 \tag{6.7a,b}$$

$$\phi=0 \quad \text{or} \quad \phi_{,\eta}=0 \quad \text{at} \quad \eta=0 \tag{6.7c,d}$$

$$\left(U_{,\eta}+1/2W_{,\eta}^2\right)W_{,\eta}+s^2\left(W_{,\eta}-\phi\right)=-\lambda\cos\alpha \quad \text{at} \quad \eta=1 \tag{6.8a}$$

$$U_{,\eta}+1/2W_{,\eta}^2=\lambda\sin\alpha \quad \text{at} \quad \eta=1 \tag{6.8b}$$

$$\phi=0 \quad \text{or} \quad \phi_{,\eta}=0 \quad \text{at} \quad \eta=1 \tag{6.8c,d}$$

and constraint condition of (6.3).

One may eliminate the function ϕ from (6.6b,c), and this results in the following equation:

$$\begin{aligned}&\left(1/\beta^2\right)\ddot{W}-\left(s^2+1\right)\ddot{W}_{,\eta\eta}+\beta^2s^2W_{,\eta\eta\eta\eta}+s^2R^2\ddot{W}+\beta^2\left[\left(U_{,\eta}+1/2W_{,\eta}\right)W_{,\eta}\right]_{,\eta\eta\eta}\\&-\left[\left(U_{,\eta}+1/2W_{,\eta}^2\right)W_{,\eta}\right]_{,\eta}^{\cdot\cdot}-s^2R^2\beta^2\left[\left(U_{,\eta}+1/2W_{,\eta}^2\right)W_{,\eta}\right]_{,\eta}=0\end{aligned} \tag{6.9}$$

The ϕ's involved in the boundary conditions may also be expressed in terms of functions W and U.

The exact solutions of the non-linear (6.6a and 6.9) can hardly be obtained. For the purpose of the present study, these two equations may be linearized by expressing the two displacement functions as power series in the amplitude, ξ, of the vibration which is assumed to be small in comparison to any dimensions of the bar. Thus, let

$$U=\xi U_1+\xi^2U_2+\ldots \qquad W=\xi W_1+\xi^2W_2+\ldots \tag{6.10a,b}$$

be substituted into (6.6a and 6.9). For small vibrations, the equations associated with the first order terms in ξ are sufficient. They are

$$\ddot{U}_1-\beta^2U_{1,\eta\eta}=0 \tag{6.11a}$$

$$W_1-\beta^2\left(1+s^2\right)\ddot{W}_{1,\eta\eta}+s^2\beta^4W_{1,\eta\eta\eta\eta}+R^2s^2\beta^2\ddot{W}_1=0 \tag{6.11b}$$

The latter is known as the equation of Timoshenko's beam theory [47], of which the general solution has been discussed in [4, 28].

Let

$$U_1 = u(\eta)\sin pt \quad W_1 = w(\eta)\sin pt \tag{6.12a,b}$$

Equations (6.11) become

$$u'' + \left(k^4/R^2\right)u = 0 \tag{6.13a}$$

$$w'''' + \left(k^4/R^2\right)\left(1 + 1/s^2\right)w'' + k^4\left[\left(k^4/R^4s^2\right) - 1\right]w = 0 \tag{6.13b}$$

with the following boundary conditions:

$$u = 0 \quad w = 0 \quad w' = 0 \quad \text{or} \quad w'' = 0 \quad \text{at} \quad \eta = 0 \tag{6.14a-d}$$

$$u' + \lambda\sin\alpha = 0 \quad w''' - R^2\lambda\cos\alpha = 0 \tag{6.15a,b}$$

$$w' = 0 \quad \text{or} \quad w'' = 0 \quad \text{at} \quad \eta = 1 \tag{6.15c,d}$$

6.2 Solutions

The solution of (6.13a) satisfying conditions (6.14a and 6.15a) is

$$u(\eta) = -\left(\lambda R/k^2\right)\sin\alpha\sin\left(k^2/R\right)\eta/\cos\left(k^2/R\right) \tag{6.16}$$

Neglecting the effect of shear deformation and including the rotatory inertia alone simplifies (6.13b) to

$$w'''' + (k^4/R^2)w'' - k^4w = 0 \tag{6.17}$$

which is also given in [33]. The general solution of both (6.17 and 6.13b) may be presented as

$$w(\eta) = c_1\sin k_1\eta + c_2\cos k_1\eta + c_3\sinh k_2\eta + c_4\cosh k_2\eta \tag{6.18}$$

in which k_1 and k_2 are characteristic values obtained from (6.17 and 6.13b), respectively,

$$k_{\frac{1}{2}} = \left(k^2/R\right)\left\{\pm 1/2 + \left[1/4 + (R/k)^4\right]^{1/2}\right\}^{1/2} \tag{6.19}$$

$$k_{\frac{1}{2}} = \left(k^2/R\right)\left\{\pm 1/2\left(1 + 1/s^2\right) + \left[1/4\left(1 + 1/s^2\right)^2 + (R/k)^4 - \left(1/s^2\right)\right]^{1/2}\right\}^{1/2} \tag{6.20}$$

Thus both k_1 and k_2 are functions of k which represents the frequency of vibration of the bar (see (2.16)).

The values of k_1 and k_2 given by (6.19) are always real. Hence the solution of (6.18) is valid for all values of k. However, when the effect of shear deformation is taken into consideration, (6.18) is true only for limited values of k. Equation (6.20) yields a real value of k_2 when $k < Rs^{1/2}$. For $k > Rs^{1/2}$,

k_2 is an imaginary number. Then the mode of (6.18) needs to be changed accordingly from the sine and cosine functions to corresponding hyperbolic functions.

This change in frequencies was first observed in [51] and was considered as a second spectrum of frequencies for the free vibration of Timoshenko's beam. This phenomenon has been discussed in a number of papers [32]. It was pointed out in [32] that such a second spectrum is merely due to the fact that when the frequencies increase to a certain value, one of the characteristic values will change from real to imaginary. The frequencies, however, are still in sequence. Thus, there is no second spectrum.

The discussions in [32, 51] were for simply supported beams. It is clear from (6.20) that there is a frequency change from real to imaginary values regardless of end conditions; the change is built in the characteristic equation.

The present work is aimed at the comparison of the coupling effect on the frequencies of free vibration of an Euler-Bernoulli inclined bar as discussed in Chap. 2 with Timoshenko's beam for inclined bars for frequencies up to those available in that chapter.

The roots of (6.20) will result in imaginary only for large k values. It is found that the k values computed in Chap. 2 are so low that no imaginary values will ever happen. Hence no further discussion about such change is necessary for the present investigation.

The four constants of integration, c_1, c_2, c_3 and c_4 in (6.18), are to be determined by the end conditions of the bars considered in what follows.

6.2.1 Inclined Bars with Hinged – "Hinged" Ends

For an inclined bar with the bottom hinged and the top end free to rotate and moveable in the vertical direction, the solution of (6.18) satisfying conditions (14b,d and 15b,d) assumes the form

$$w(\eta) = \frac{\lambda R^2 \cos\alpha}{k_1^2 k_2^2} \frac{k_2^2 \sinh k_2 \sin k_1\eta + k_1^2 \sin k_1 \sinh k_2\eta}{k_2 \sin k_1 \cosh k_2 - k_1 \cos k_1 \sinh k_2} \tag{6.21}$$

Substitution of (6.16 and 6.21) into the constraint condition (6.3), for $\lambda \neq 0$, yields the frequency equation

$$\tan^2\alpha - k^2 R \cot\left(k^2/R\right)\left(\frac{1}{k_1^2} + \frac{1}{k_2^2}\right) \frac{\sin k_1 \sinh k_2}{k_2 \cosh k_2 \sin k_1 - k_1 \sinh k_2 \cos k_1} = 0 \tag{6.22}$$

When $\alpha = 0$, (6.22) calls for

$$\cos\left(k^2/R\right) = 0 \quad \text{then} \quad k^2/R = \pi/2, 3\pi/2, 5\pi/2, \ldots \tag{6.23a}$$

which result in

$$k = \sqrt{\pi R/2}, \sqrt{3\pi R/2}, \sqrt{5\pi R/2}, \ldots; \quad \text{for} \quad R = 50, k = 8.862, 12.350\ldots. \tag{6.23b}$$

Table 6.1. First six frequencies $R = 50,\ S^2 = 1/3$

α	Eqs.	First	Second	Third	Fourth	Fifth	Sixth
			Hinged-"hinged" bars				
0°	E-B	3.142	6.283	8.862	9.425	12.566	15.350
	(17)	3.146	6.257	8.862	9.336	12.373	15.350
	(13b)	3.127	6.188	8.862	9.116	11.901	14.510
90°	E-B	0	3.927	7.069	10.210	12.533	13.352
	(17)	0	3.919	7.036	10.115	12.533	13.134
	(13b)	0	3.908	6.951	9.866	12.533	12.610
			Clamped-"clamped" bars				
0°	E-B	4.730	7.583	8.862	10.996	14.137	15.708
	(17)	4.727	7.818	8.862	10.895	13.923	15.708
	(13b)	4.707	7.718	8.862	10.611	13.345	15.708
90°	E-B	2.365	5.498	8.639	11.781	12.533	14.923
	(17)	2.362	4.482	8.583	11.644	12.533	14.649
	(13b)	2.359	5.442	8.428	11.264	12.533	13.935

These values are for the axial longitudinal vibrations, thus they are not affected by end conditions of the beams as shown in Tables 6.1 and also in Table 6.2. For $\alpha = 0$, $k_1 \neq 0$, (6.22) also is satisfied by

$$\sin k_1 = 0 \quad \text{then} \quad k_1 = \pi, 2\pi, 3\pi, \ldots \tag{6.24}$$

When $\alpha = \pi/2$, one requires

$$\sin\left(k^2/R\right) = 0 \quad \text{then} \quad k^2/R = \pi, 2\pi, 3\pi, \ldots \tag{6.25a}$$

or

$$k = \sqrt{\pi R},\ \sqrt{2\pi R},\ \sqrt{3\pi R}\,, \ldots; \quad \text{for} \quad R = 50, k = 12.533, 17.722, \ldots \tag{6.25b}$$

For $\alpha = \pi/2$, (6.22) also calls for

$$k_2 \tan k_1 = k_1 \tanh k_2 \tag{6.26}$$

For a bar with $R = 50$ and $s^2 = 1/3$, the first seven sets of curves for the frequency k versus the angle α as determined from (6.22) are presented in Fig. 6.1.

In Figs. 6.1 to 6.4, the dash-dot curves labeled "R" are obtained from (6.17) including rotatory inertia alone, and the solid lines, "R and S", are obtained from (6.13b) including both rotatory inertia and shear deformation. The labels "R" and "S" and the parameters of slenderness ratio, R, and shear coefficient, s, should not be confused.

Table 6.2. First six frequencies $R = 50$, $S^2 = 1/3$

α	Eqs.	First	Second	Third	Fourth	Fifth	Sixth
Bottom hinged-top "clamped" bars							
0°	E-B	3.927	7.089	8.862	10.210	13.352	15.350
	(17)	3.920	7.037	8.862	10.114	13.138	15.350
	(13b)	3.908	6.952	8.862	9.863	12.621	15.350
90°	E-B	1.875	4.694	7.855	10.996	12.533	14.137
	(17)	1.867	4.685	7.817	10.828	12.533	13.904
	(13b)	1.867	4.664	7.717	10.612	12.533	13.339
Bottom clamped-top "hinged" bars							
0°	E-B	3.927	7.089	8.862	10.210	13.352	15.350
	(17)	3.920	7.037	8.862	10.114	13.138	15.350
	(13b)	3.908	6.951	8.862	9.863	12.621	15.350
90°	E-B	1.571	4.712	7.854	10.996	12.533	14.137
	(17)	1.563	4.702	7.806	10.865	12.533	13.864
	(13b)	1.561	4.668	7.672	10.531	12.533	13.222

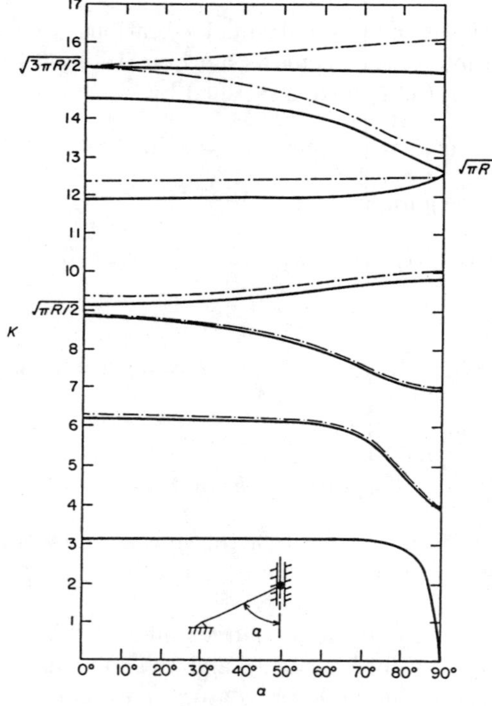

Fig. 6.1. Frequencies of hinged-"hinged" bars — · —, R; —, R & S

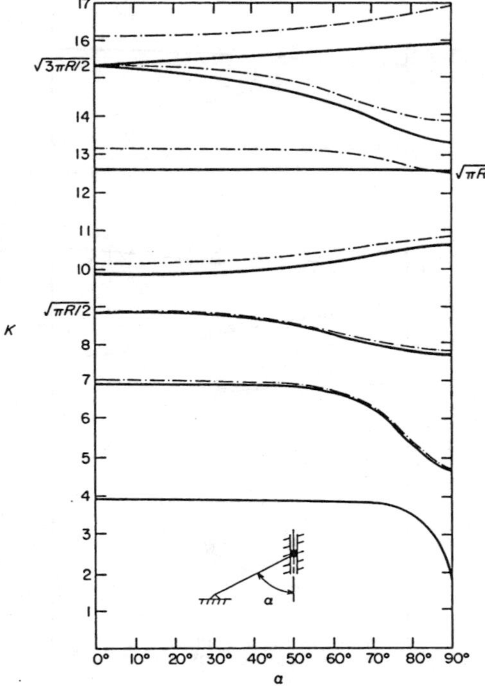

Fig. 6.2. Frequencies of hinged-"clamped" bars — · —, R; —, R & S

The first six frequencies for $\alpha = 0$ and $\pi/2$ are listed in Table 6.1. For comparison, frequencies obtained in Chap. 2 where the Euler-Bernoulli (E-B) equation was used are also listed on the first line in Tables 6.1 and 6.2 in each case.

Equation (6.23 and 6.25) for the axial longitudinal vibration are true for all bars considered.

6.2.2 Inclined Bars with Bottom Hinged-Top "Clamped"

When the boundary conditions of (6.14b,d and 6.15b,c) are satisfied, (6.18) yields

$$w(\eta) = -\frac{\lambda R^2 \cos\alpha}{k_1 k_2 \left(k_1^2 + k_2^2\right)} \frac{k_2 \cosh k_2 \sin k_1 \eta - k_1 \cos k_1 \sinh k_2 \eta}{\cos k_1 \cosh k_2} \tag{6.27}$$

The frequency equation is

$$\tan^2\alpha - k^2 R \cot\left(k^2/R\right) \frac{k_1 \tanh k_2 - k_2 \tan k_1}{k_1 k_2 \left(k_1^2 + k_2^2\right)} = 0 \tag{6.28}$$

which, for $\alpha = 0$ and $\pi/2$, reduces to, respectively,

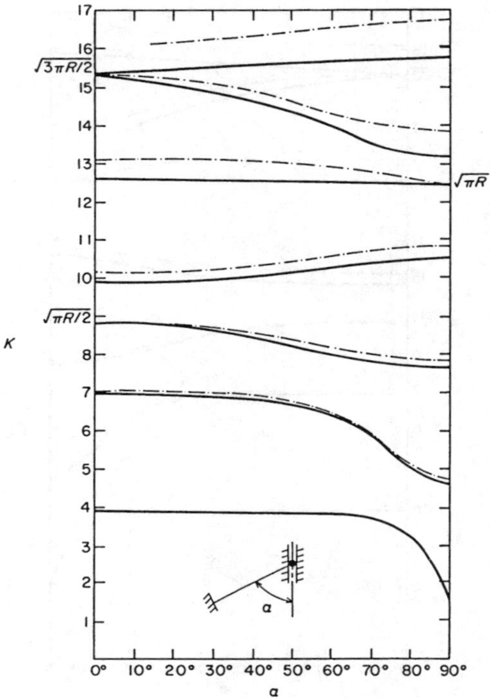

Fig. 6.3. Frequencies of clamped-“hinged” bars — · —, R; —, R & S

$$K_2 \tan k_1 = K_1 \tanh K_2 \tag{6.29}$$

$$\tan k_1 = \infty \quad K_1 = \pi/2, 3\pi/2, 5\pi/2, \ldots \tag{6.30}$$

The first seven frequencies are shown in Fig. 6.2 and the first six frequencies when $\alpha = 0$ and $\pi/2$ are listed in Table 6.2.

6.2.3 Inclined Bars with Bottom Clamped-Top “Hinged”

When conditions (6.14b,c and 6.15b,d) are satisfied, the general solution (6.18) becomes

$$\begin{aligned} w(\eta) = -\frac{\lambda R^2 \cos\alpha}{B} & \left[\left(k_1^2 \cos k_1 + k_2^2 \cosh k_2\right)\left(k_2 \sin k_1\eta - k_1 \sinh k_2\eta\right)(1/k_2)\right. \\ & \left. - \left(k_1^2 \sin k_1 + k_1 k_2 \sinh k_2\right)\left(\cos k_1\eta - \cosh k_2\eta\right)\right] \end{aligned} \tag{6.31}$$

in which

$$B = k_1^5 + k_1 k_2^4 + 2k_1^3 k_2^2 \cos k_1 \cosh k_2 + k_1^2 k_2 \left(k_1^2 - k_2^2\right) \sin k_1 \sinh k_2 \tag{6.32}$$

The frequency equation obtained from condition (6.3) for $\lambda \neq 0$ is

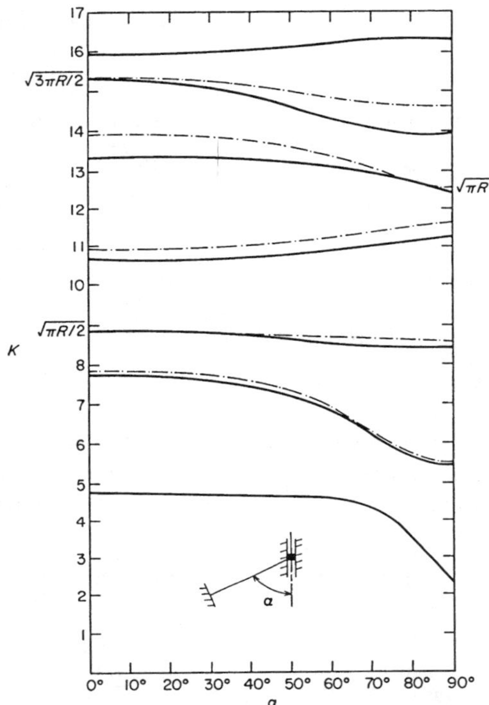

Fig. 6.4. Frequencies of clamped-"clamped" bars — · —, R; —, R & S

$$\tan^2 \alpha - Rk^2 \cot\left(k^2 R\right)\left(k_1^2 + k_2^2\right)\left[k_1 \cos k_1 \sinh k_2 - k_2 \sin k_1 \cosh k_2\right]/k_2 B = 0 \tag{6.33}$$

which, for $\alpha = 0$, reduces to (6.29); for $\alpha = \pi/2$ and $k_1 \neq 0$, from (6.31) or (6.32) one has

$$k_1^4 + k_2^4 + 2k_1^2 k_2^2 \cos k_1 \cosh k_2 + k_1 k_2 \left(k_1^2 - k_2^2\right) \sin k_1 \sinh k_2 = 0 \tag{6.34}$$

The results are presented in Fig. 6.3 and Table 6.2.

6.2.4 Inclined Bars with Bottom Clamped-Top Clamped in Rotation

With conditions (6.14b,c and 6.15b,c) satisfied, the general solution (6.18) becomes

$$\begin{aligned} w(\eta) = {} & -\lambda R^2 \cos\alpha \left[\left(k_2 \sin k_1 \eta - k_1 \sinh k_2 \eta\right)\left(k_1 \sin k_1 + k_2 \sinh k_2\right)\right. \\ & \left. + k_1 k_2 \left(\cos k_1 \eta - \cosh k_2 \eta\right)\left(\cos k_1 - \cosh k_2\right)\right]/k_1 k_2 D \end{aligned} \tag{6.35}$$

where

$$D = \left(k_1^2 + k_2^2\right)\left(k_1 \sin k_1 \cosh k_2 + k_2 \cos k_1 \sinh k_2\right) \tag{6.36}$$

The frequency equation is

$$\tan^2 \alpha - Rk^2 \cot\left(k^2/R\right)\left[\left(k_1^2 - k_2^2\right) \sin k_1 \sinh k_2/k_1 k_2 + 2\cos k_1 \cosh k_2 - 2\right]/D = 0 \tag{6.37}$$

For $\alpha = 0$ and $\alpha = \pi/2$, (6.37) reduces to, respectively,

$$\left(k_1^2 - k_2^2\right) \sin k_1 \sinh k_2/k_1 k_2 + 2\cos k_1 \cosh k_2 - 2 = 0 \tag{6.38}$$

$$k_1 \tan k_1 + k_2 \tanh k_2 = 0 \tag{6.39}$$

The frequencies are shown in Fig. 6.4 and Table 6.1.

In order to show the effect of slenderness ratio on the frequencies, when both rotatory inertia and shear deformation are taken into consideration, for the present case of clamped-"clamped" bars, frequencies for $\alpha = 0$ and $\pi/2$ are presented along the two edges in Fig. 6.5, for $R = 50, 100, 200$.

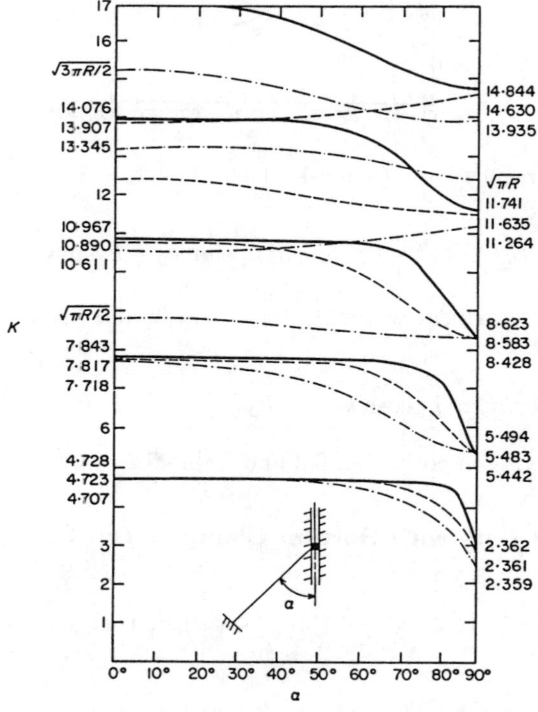

Fig. 6.5. Frequencies of clamped-"clamped" bars —, $R = 200$; ······. $R = 100$; — · —, $R = 50$

It is shown in Tables 6.1 and 6.2 that neither shear deformation and rotatory inertia nor slenderness ratio have effect on the axial vibration frequencies. For example, the third frequency $k = (\pi R/2)^{1/2} = 8.862$ remains for all cases.

In Figs. 6.1 to 6.4, there is no difference for the first frequencies for (6.13b and 6.17). As the frequencies get higher, the differences become apparent, especially after the frequency of axial vibration $k = (\pi R/2)^{1/2}$. It is also seen from Fig. 6.5 that after the frequency of axial vibration, the coupling effect makes the frequency spectra much more complicated than the simple beam vibration.

The effect of rotatory inertia and shear deformation on the normal mode curves was also examined for the case of hinged-"hinged" and clamped-"clamped" bars. It was seen that except for the slight changes of the amplitudes up to the sixth mode, these effects were not much different from those presented in Chap. 2. Hence these curves are not presented here.

Exercises

1. From (6.4), derive three (6.6) and boundary conditions (6.7 and 6.8).
2. Verify characteristic (6.19 and 6.20).
3. Verify the frequency of (6.22). And determine the frequencies for two limit cases of $\alpha = 0$ and $\pi/2$ from this equation.

7

Out-of-Plane Vibrations of Plane Frames

The frame vibrations studied in Chaps. 3 to 5 are for in-plane vibrations. If a plane frame acted upon by a force or horizontal ground vibration in the perpendicular direction of the plane, the frame will experience out-of-plane vibrations as shown in a numerical example in [41].

Free and forced out-of-plane vibrations of elastic plane frames are investigated in the present chapter. Analytic solutions of Euler-Bernoulli equation coupled with axial and torsional vibrations are given for portal frames without and with x-braces.

7.1 Formulation

Consider an elastic portal frame as shown in Fig. 7.1, consisting of i ($i = 1, 2, 3$, presently) bar elements which may not have the same geometrical or material properties. The ratio of the beam length to the column length is defined by the angle α. In general, the ith bar, shown in Fig. 7.2, may be undergoing a longitudinal vibration $\bar{U}_i$, two transverse vibrations $\bar{V}_i$ and $\bar{W}_i$ and a torsional rotation $\bar{\theta}_i$, all of which are functions of local co-ordinates, x_i, and time, t. It was shown in Chap. 2 that the equations of motion for these functions can be derived from Hamilton's theory.

For small vibration, these equations may be linearized and separated. They are in the following forms:

$$E_i \frac{\partial^2 \bar{U}_i}{\partial x_i^2} - \rho_i \frac{\partial^2 \bar{U}_i}{\partial t^2} = 0 \quad E_i I_{zi} \frac{\partial^4 \bar{V}_i}{\partial x_i^4} + \rho_i \bar{A}_i \frac{\partial^2 \bar{V}_i}{\partial t^2} = 0 \tag{7.1a,b}$$

$$E_i I_{yi} \frac{\partial^4 \bar{W}}{\partial x_i^4} + \rho_i \bar{A}_i \frac{\partial^2 \bar{W}_i}{\partial t^2} = 0 \quad G_i \frac{\partial^2 \bar{\theta}_i}{\partial x_i^2} - \rho_i \frac{\partial^2 \bar{\theta}_i}{\partial t^2} = 0 \tag{7.1c,d}$$

in which $E_i, I_{yi}, I_{zi}, \rho_i, \bar{A}_i$, and G_i are the modulus of elasticity, the area moments of inertia about the local axes, y and z, the mass density, the cross-sectional area, and the shear modulus of the ith bar, respectively.

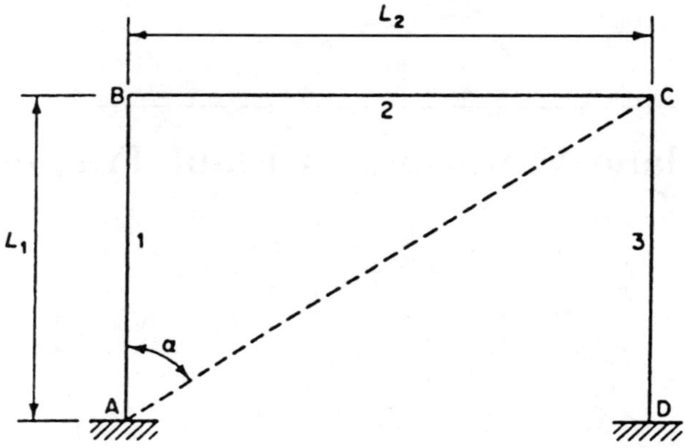

Fig. 7.1. A portal frame

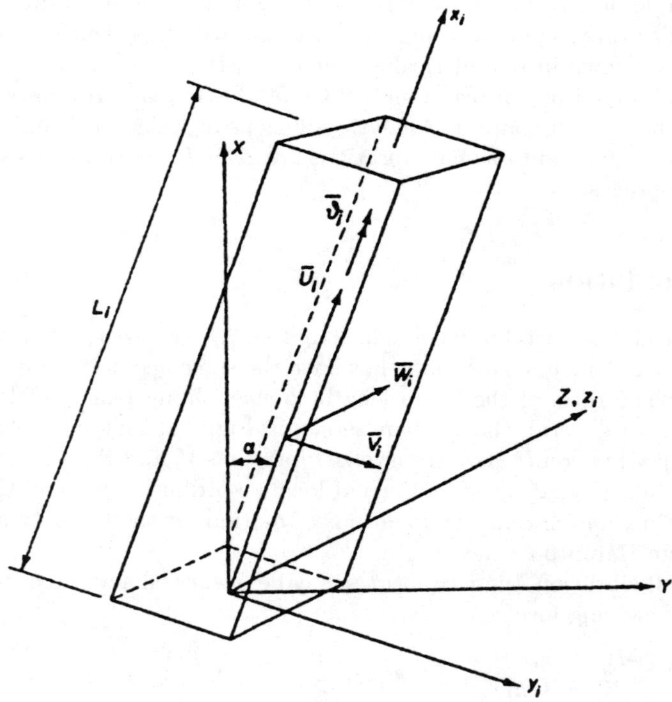

Fig. 7.2. A bar element

When the frame is undergoing in-plane vibration in the x-y plane, the transverse vibration $\bar{V}_i$ is coupled with the axial vibration $\bar{U}_i$, as shown in Chaps. 2 and 3. In out-of-plane vibration, they are coupled in the x-z plane, when the displacements $\bar{W}_1$ and $\bar{W}_3$ in the two columns are in-phase for identical column modes. The two columns behave as two cantilever beams, and the horizontal central beam will experience a rigid body motion. But when $\bar{W}_1$ and $\bar{W}_3$ are in an out-of-phase mode, a torsional vibration, $\bar{\theta}_i$, will be developed in the central beam as well as in the two columns if the two bottom ends of the two columns are fixed and the top two joints are rigidly connected. In this case, there are six boundary conditions at the bottom end A at $x_1 = 0$ in Fig. 7.1:

$$\bar{U}_1 = \bar{V}_1 = \frac{\partial \bar{V}_1}{\partial x_1} = 0 \quad \bar{W}_1 = \bar{\theta}_1 = \frac{\partial \bar{W}_1}{\partial x_1} = 0 \tag{7.2a-f}$$

At joint B, $x_1 = L_1, x_2 = 0$, there are six conditions of continuity:

$$\bar{U}_1 = -\bar{V}_2 \quad \bar{V}_1 = \bar{U}_2 \quad \frac{\partial \bar{V}_1}{\partial x_1} = \frac{\partial \bar{V}_2}{\partial x_2} \tag{7.3a-c}$$

$$\bar{W}_1 = \bar{W}_2 \quad \bar{\theta}_1 = \frac{\partial \bar{W}_2}{\partial x_2} \quad \frac{\partial \bar{W}_1}{\partial x_1} = -\bar{\theta}_2 \tag{7.3d-f}$$

and six equations of equilibrium:

$$\bar{A}_1 E_1 \frac{\partial \bar{U}_1}{\partial x_1} = E_2 I_{z2} \frac{\partial^3 \bar{V}_2}{\partial x_2^3} \quad \bar{A}_2 E_2 \frac{\partial \bar{U}_2}{\partial x_2} = -E_1 I_{z1} \frac{\partial^3 \bar{V}_1}{\partial x_1^3}$$
$$E_1 I_{z1} \frac{\partial^2 \bar{V}_1}{\partial x_1^2} = E_2 I_{z2} \frac{\partial^2 \bar{V}_2}{\partial x_2^2} \tag{7.4a-c}$$

$$E_1 I_{y1} \frac{\partial^3 \bar{W}_1}{\partial x_1^3} = E_2 I_{y2} \frac{\partial^3 \bar{W}_2}{\partial x_2^3} \quad G_1 J_1 \frac{\partial \bar{\theta}_1}{\partial x_1} = E_2 I_{y2} \frac{\partial^2 \bar{W}_2}{\partial x_2^2}$$
$$G_2 J_2 \frac{\partial \bar{\theta}_1}{\partial x_2} = -E_1 I_{y1} \frac{\partial^2 \bar{W}_1}{\partial x_2^2} \tag{7.4d-f}$$

where J_i is a torsional constant which, if the bar is circular, is the polar moment of the cross-sectional area inertia. For other sections, the J value may be found from available formulas as given in [21].

Thus there are a total of eighteen conditions at joints A and B. There are another similar eighteen conditions at joints C and D. There are a total of thirty six conditions to determine thirty six constants of integration for the portal frame shown in Fig. 7.1.

The first three sets of (7.2a-c, 7.3a-c and 7.4a-c) involving $\bar{U}_i$ and $\bar{V}_i$ are for in-plane vibrations, while the other sets for $\bar{W}_i$ and $\bar{\theta}_i$ are for out-of-plane vibrations. Therefore, these two types of vibrations are separable. The solutions of (7.1c,d, 7.2d-f, 7.3d-f and 7.4d-f) are for out-of-plane vibrations and they are considered here.

Let

$$\bar{W}_i = W_i(\eta_i) L_1 \cos(pt) \quad \bar{\theta}_i = \theta_i(\eta_i) \cos(pt) \tag{7.5a,b}$$

in which p is the circular frequency and

$$\eta_i = x_i / L_1 \tag{7.6}$$

Equations (7.1c,d) now may be reduced to the dimensionless ordinary differential equations

$$W_i'''' - K^4 m_i^4 S_i^4 W_i = 0 \quad \theta_i'' + \frac{\mu_i^2 K^4 m_i^4}{R^2} \theta_i = 0 \tag{7.7a,b}$$

in which a prime indicates a derivative with respect to η_i and

$$\begin{aligned} K = \left(\rho_1 L_1^2 R^2 p^2 / E_1\right)^{1/4} &= \text{ frame frequency parameter} \\ m_i = \left(E_1 \rho_i / E_i \rho_1\right)^{1/4} &= \text{material constant} \\ S_i = \left(r_1 / r_i\right)^{1/2} &= \text{stiffness factor} \\ R = L_1 / r_1 &= \text{ slenderness ratio of bar 1} \\ \mu_i = \left(E_i / G_i\right)^{1/2} = \sqrt{2}\left(1 + \nu_i\right)^{1/2} &= \text{ equivenlet Poisson's ratio} \\ r_i = \left(I_{yi} / \bar{A}_{\ i}\right)^{1/2} &= \text{ radius of gyration} \\ A_{\ i} = \bar{A}_{\ i} / \bar{A}_{\ 1} &= \text{ area ration} \end{aligned} \tag{7.8a-g}$$

where ν_i is the Poisson's ratio. Then, the solutions of (7.7a,b) are, respectively,

$$\begin{aligned} W_i = C_{i1} \sin(K m_i S_i \eta_i) + C_{i2} \cos(K m_i S_i \eta_i) \\ + C_{i3} \sinh(K m_i S_i \eta_i) + C_{i4} \cosh(K m_i S_i \eta_i) \end{aligned} \tag{7.9a}$$

$$\theta_i = C_{i5} \sin\left(\mu_i K^2 m_i^2 \eta_i / R\right) + C_{i6} \cos\left(\mu_i K^2 m_i^2 \eta_i / R\right) \tag{7.9b}$$

where C_{is} $(s = 1, 2, \ldots, 6)$ are constants of integration.

The shear, moment and torque are

$$\Phi_i = \frac{\bar{A}_i E_i}{R^2 S_i^4} W_i''' \quad M_i = \frac{E_i I_i}{L_1} W_i'' \quad T_i = \frac{G_i J_i}{L_1} \theta_i' \tag{7.10a-c}$$

The positive directions of these forces and moments are shown in Fig. 7.3.

7.2 Free Vibration

For the frame shown in Fig. 7.1, there are three bars and each has a set of solutions of (7.9). Therefore, there are 18 constants of C_{is} $(i = 1, 2, 3$ and $s = 1, 2, \ldots, 6)$. As stated earlier, this frame has 6 boundary conditions at the two bottom ends and 12 conditions at the two top joints. These 18 homogeneous equations are used to determine the 18 unknown constants.

For an x-braced portal frame shown in Fig. 7.4, there are another 12 constants in the solutions for the two braces. Since the angles of twist due to

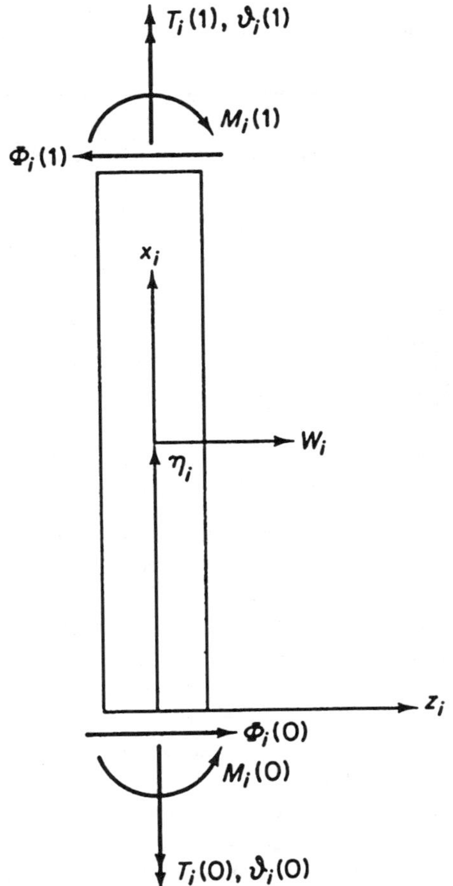

Fig. 7.3. Forces and moments on an element

torsion and bending are small for the linearized small vibration, they may be treated as vectors. Thus, the angles from the diagonal braces may be resolved into two components acting along the local Cartesian coordinates to obtain the required additional 12 conditions. Totally, there are 30 conditions to determine the 30 unknown constants C_{is} $(i = 1, 2, \ldots, 5)$for an x-braced frame.

Let $C_j = C_{is}$, where $(j = 1, \ldots, 18$, for the portal frame, and $j = 1, \ldots, 30$, for the braced frame). These homogeneous equations may be expressed in matrix form as

$$[A_{ij}]\,[C_j] = 0 \tag{7.11}$$

The coefficient functional matrix $[A]$ contains the natural frequency parameter K. For the non-trivial solution of C_j, the determinant of the matrix

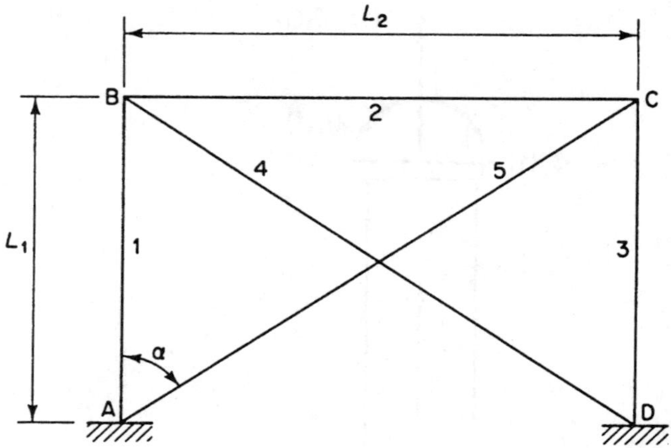

Fig. 7.4. An x-braced portal frame

$[A]$ must vanish. This yields the eigenvalues or frame frequency K values. For each value of K, the eigenvector C_j and hence the corresponding mode may be determined.

The first five K values for the portal frame with $R = 50$, 100 and 200 for different α's of identical bars, where $A_2 = 1$ and $S_2 = 1$, are presented in Table 7.1. For the portal frames with heavy beams, where $A_2 = 10$ and $S_2 = 0.5$, the K values are given in Table 7.2. The first five modes, for $\alpha = 45°$, $R = 100$, $A_2 = 1$ and $S_2 = 1$, are shown in Fig. 7.5, while for$A_2 = 10$ and $S_2 = 0.5$, they are depicted in Fig. 7.6.

The first five K values of the x-braced frames with identical bars, $A_i = 1.0$ and $S_i = 1.0$, are given in Table 7.3. The results indicate that there is little change for different Rs. For frames with a heavy horizontal beam, $A_2 = 10$, R has no effect to the frequencies at all. For different braces, $A_4 = 1.0, 0.5, 0.1$, the K values are different, and they are presented in Tables 7.4, 7.5, and 7.6, respectively. The first five modes for $\alpha = 45°$, $R = 100$ and $A_i = 1.0$ are depicted in Fig. 7.7; those for $A_2 = 10$ and $A_4 = 0.5$ are shown in Fig. 7.8. The heavy beam remains straight for all modes presented in Fig. 7.8.

7.3 Forced Vibration

In this section, the vibration responses of a portal frame and an x-braced frame due to both sinusoidal time dependent forces and horizontal ground motion are studied.

For forced steady vibration, the case is considered where two equal concentrated forces are applied at the top joints B and C in direction perpendicular to the plane of the frames. These two forces act in the same direction. Let

$$F = f\bar{A}_1 E_1 \cos(\lambda^2 pt) \tag{7.12}$$

Table 7.1. First five frequencies of portal frames $A_i = 1.0,\ S_i = 1.0$

α (degrees)				
15	30	45	60	75
$R = 50$				
1.682	1.543	1.413	1.250	0.890
3.264	2.482	2.070	1.756	1.357
4.334	4.139	3.702	2.597	1.620
5.195	4.778	4.490	3.875	2.165
7.332	6.498	4.768	4.233	2.846
$R = 100$				
1.683	1.543	1.414	1.251	0.890
3.279	2.485	2.071	1.756	1.358
4.340	4.148	3.709	2.559	1.621
5.233	4.784	4.494	3.880	2.166
7.360	6.573	4.803	4.267	2.848
$R = 200$				
1.683	1.543	1.414	1.251	0.890
3.282	2.486	2.071	1.756	1.358
4.341	4.151	3.711	2.600	1.622
5.242	4.785	4.495	3.881	2.166
7.366	6.589	4.811	4.275	2.848

Table 7.2. First five frequencies for portal frames $A_2 = 10.0,\ S_2 = 0.5$

α (degrees)				
15	30	45	60	75
$R = 50$				
1.168	0.984	0.864	0.756	0.624
2.847	2.001	1.648	1.408	1.153
3.415	3.019	2.709	2.401	1.978
5.039	4.889	4.817	4.707	2.533
5.354	5.040	4.899	4.752	4.095
$R = 100$				
1.173	0.988	0.868	0.760	.0628
2.856	2.002	1.648	1.408	1.153
3.840	3.643	3.439	3.171	2.491
5.045	4.889	4.817	4.756	2.752
6.242	5.690	5.335	4.938	4.097
$R = 200$				
1.174	0.990	0.870	0.761	0.628
2.859	2.002	1.648	1.408	1.153
3.669	3.883	3.810	3.700	2.512
5.046	4.890	4.817	4.757	3.467
6.890	6.609	6.285	5.348	4.097

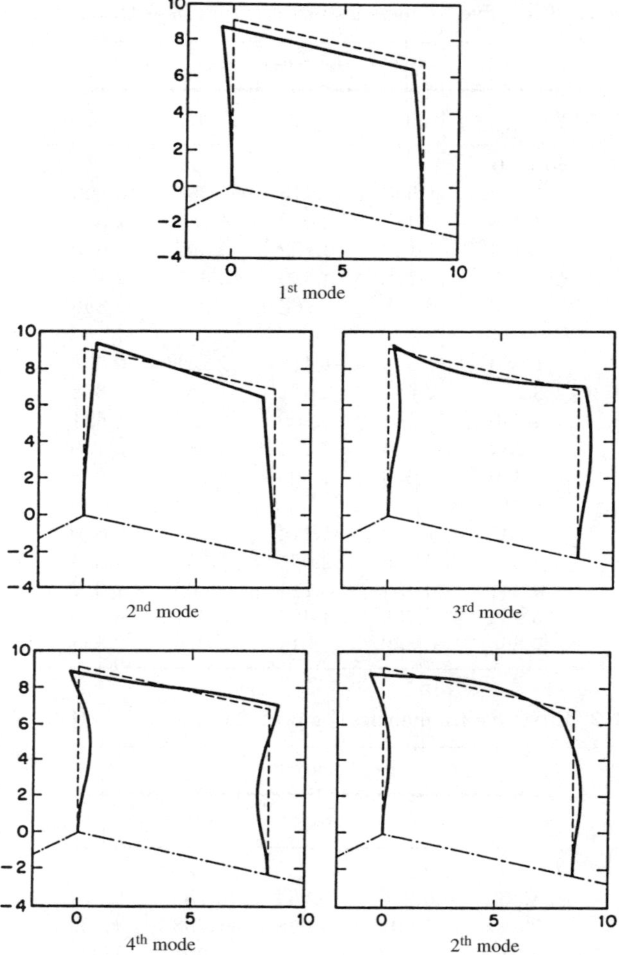

Fig. 7.5. First five modes of a portal frame with $\alpha = 45°$, $R = 100$ and $A_i = S_i = 1$

and

$$\bar{w}_i = w_i(\eta_i)L_1\cos(\lambda^2 pt) \quad \bar{\theta}_i = \theta_i(\eta_i)\cos(\lambda^2 pt) \tag{7.13a,b}$$

where f is a dimensionless force and λ is a frequency parameter, hence $\lambda^2 p$ represents the frequency of the external force. The solutions of (7.9) are still valid by simply replacing K by

k, such that $$k = \lambda K \qquad 0 \le \lambda < 1 \tag{7.14}$$

Then

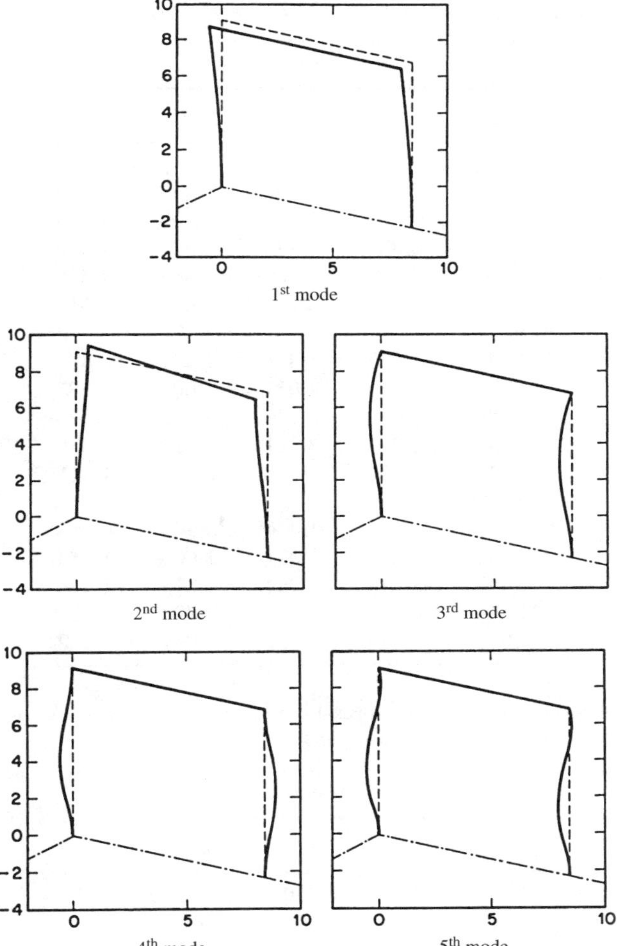

Fig. 7.6. First five modes of a portal frame with $\alpha = 45°$, $R = 100$ and $A_2 = 10$, $S_2 = 0.5$

$$\begin{aligned} w_i &= c_{i1}\sin(km_iS_i\eta_i) + c_{i2}\cos(km_iS_i\eta_i) \\ &\quad + c_{i3}\sinh(km_iS_i\eta_i) + c_{i4}\cosh(km_iS_i\eta_i) \qquad (7.15a) \\ \theta_i &= c_{i5}\sin(\mu_ik^2m_i^2\eta_i/R) + c_{i6}\cos(\mu_ik^2m_i^2\eta_i/R) \qquad (7.15b) \end{aligned}$$

The frames have the same boundary and joint conditions as in free vibrations, except that for the equations of equilibrium at joints B and C, (7.4d) now becomes

$$E_1I_{y1}\bar{w}_1''' - E_2I_{y2}\bar{w}_2''' = -f\bar{A}_1E_1 \qquad (7.16a)$$

or, in dimensionless form,

Table 7.3. First five frequencies for x-braced frames $A_2 = 1.0,\ S_2 = 1.0$

α (degrees)				
15	30	45	60	75
$R = 50$				
1.745	1.599	1.435	1.240	0.885
3.314	2.452	1.970	1.600	1.085
4.382	3.938	3.242	2.311	1.223
4.537	4.009	3.260	2.322	1.355
4.643	4.367	3.763	2.649	1.511
$R = 100$				
1.745	1.599	1.436	1.241	0.885
3.329	2.455	1.971	1.600	1.085
4.386	3.942	3.243	2.312	1.223
4.541	4.011	3.261	2.322	1.355
4.644	4.377	3.778	2.650	1.511
$R = 200$				
1.745	1.599	1.436	1.241	0.885
3.333	2.455	1.971	1.600	1.085
4.387	3.943	3.244	2.312	1.224
4.543	4.011	3.261	2.323	1.355
4.644	4.380	3.781	2.651	1.512

Table 7.4. First five frequencies for x-braced frames $A_2 = 10.0,\ S_2 = 0.5,\ A_4 = 1.0,\ S_4 = 1.0$

α (degrees)				
15	30	45	60	75
1.336	1.119	0.956	0.808	0.644
3.102	2.095	1.659	1.384	1.107
4.005	3.690	3.151	2.315	1.224
4.597	4.131	3.370	2.383	1.250
4.653	4.458	4.216	3.747	2.024

$$\frac{1}{R^2}\left(w_1''' - \frac{E_2}{E_1}\frac{A_2}{S_2^4}w_2'''\right) = -f \tag{7.16b}$$

Then the homogeneous (7.11) now has the form

$$[A_{ij}]\,[c_j] = [b_j] \tag{7.17}$$

in which $b_j = 0$ except at joints B and C, where $b_j = -f$.

For a given value of λ ($\lambda \neq 1$), the inverse of the coefficient matrix exists. Thus

$$[c_j] = [A_{ij}]^{-1}\,[b_j] \tag{7.18}$$

Table 7.5. First five frequencies for x-braced frames $A_2 = 10.0$, $S_2 = 0.5$, $A_4 = 0.5$, $S_4 = 2.0$

α (degrees)				
15	30	45	60	75
1.161	0.985	0.867	0.758	0.604
2.271	1.929	1.591	1.177	0.612
2.286	2.046	1.673	1.184	0.635
2.827	2.100	1.713	1.398	1.012
3.765	3.386	2.773	1.964	1.017

Table 7.6. First five frequencies for x-braced frames $A_2 = 10.0$, $S_2 = 0.5$, $A_4 = 0.1$, $S_4 = 4.0$

α (degrees)				
15	30	45	60	75
1.132	0.984	0.834	0.761	0.628
1.142	1.024	0.836	0.981	0.711
1.182	1.029	0.872	0.982	0.712
1.896	1.699	1.387	1.370	1.117
1.898	1.701	1.389	1.375	1.118

Substitution of the solution of (7.18) into (7.15) yields the steady state solutions of forced vibration.

For the steady vibration of the frames due to horizontal sinusoidal ground vibrations, a similar approach can be employed. In this case, the conditions at all joints remain the same as in the free vibration, except for the boundary conditions at the bottom ends.

For the portal frame,

$$\bar{w}_1(0) = \bar{w}_3\left(L_3\right) = gL_1 \cos\left(\lambda^2 pt\right) \tag{7.19a,b}$$

For the x-braced frame, in addition to the above two conditions,

$$\bar{w}_4(L_4) = \bar{w}_5(0) = gL_1 \sin(\lambda^2 pt) \tag{7.20a,b}$$

in which g is a dimensionless ground vibration amplitude. With the application of these conditions, the solutions of (7.18), and hence of (7.15), are obtained for a set of given values of g and λ.

7.4 Numerical Examples

Consider a portal frame consisting of a heavy, 6.1 m (20 ft) long, S61 cm × 1751 N/m (S24 × 12), steel I-beam rigidly connected to two 3.66 m (12ft)

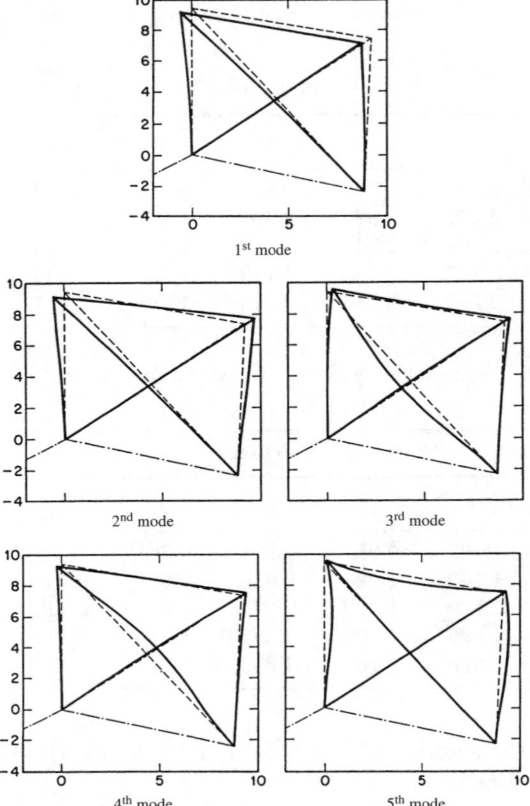

Fig. 7.7. First five modes of an x-braced frame with $\alpha = 45°$, $R = 100$ and $A_i = 1$, $S_i = 1$

long columns made of two C20.3 cm × 200.7N/m (C8 × 13.75) channels. The first five natural frequencies are presented in Table 7.7. Equation (7.8a) was used in the computation of the frequency values. This frame is braced by two diagonals, each is made of one L20.32 cm × 20.32cm × 1.72 cm (L8 × 8 × 1/2) angle. The first five frequencies are given in Table 7.8.

Let both frames be subjected to a pair of equally concentrated forces acting on joints B and C. The dynamic response amplitudes of joint B for $f\bar{A}_1E_1 = 1112\,N$ (250 lb), and for $\lambda = 0.5$, 0.7, and 0.9 are depicted in Fig. 7.9 for the portal frame, and in Fig. 7.10 for the x-braced frame. For reference, the static responses which were obtained by letting $\lambda = 0$ are also included. These two figures indicate that the response amplitudes of the x-braced frame are about 20% less than that of the portal frame in the present example.

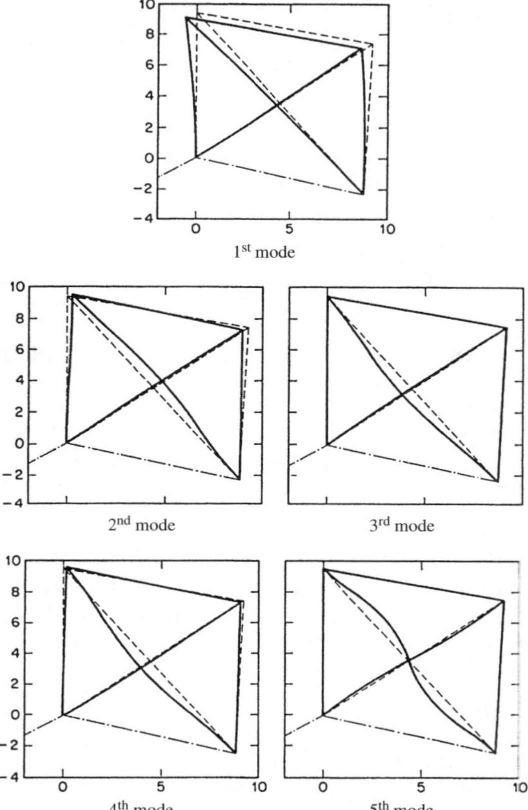

Fig. 7.8. First five modes of an x-braced frame with $\alpha = 45°$, $R = 100$ and $A_2 = 10$. $S_2 = 0.5$, $A_4 = 0.5$, $S_4 = 2.0$

The responses due to ground vibrations of $g = 0.0025$ or $gL_1 = 0.914\,\text{cm}$ (0.36 in) are presented in Fig. 7.11 for the portal frame and in Fig. 7.12 for the x-braced frame. In this case, the maximum deflections are almost the same.

7.5 Discussion

The solutions presented are exact for out-of-plane vibrations of plane frames when small vibration theory is used. The elastic torsional vibration has been taken into consideration. The torsional effect, however, is very small in the basic modes. This means that the torsional inertia represented by the second term in (7.1d) may be neglected. Its effect is appreciable only for modes higher than those presented here. In that case, the Euler-Bernoulli equation of beam vibration has to be replaced by Timoshenko's beam theory. Nevertheless, the

Table 7.7. First five frequencies of a portal frame

Number	Frequency Parameter, K	Radian Frequency p (rad/sec)	Frequency, F(1/sec)	Period T(sec)
1	7.76930E − 01	4.99527E + 00	7.95022E − 01	1.25783E + 00
2	1.05598E + 00	9.22793E + 00	1.46867E + 00	6.80888E − 00
3	2.02221E + 00	3.38404E + 01	5.38603E + 00	1.85666E − 01
4	3.26685E + 00	8.81569E + 01	1.40306E + 01	7.12728E − 02
5	3.94065E + 00	1.28508E + 02	2.04527E + 01	4.88933E − 02

Table 7.8. First five frequencies of an x-braced frame

Number	Frequency Parameter, K	Radian Frequency p (rad/sec)	Frequency, F(1/sec)	Period T(sec)
1	7.92898E − 01	5.20272E + 00	8.28039E − 01	1.20767E + 00
2	1.13802E + 00	1.07176E + 01	1.70576E + 00	5.86249E − 01
3	2.09299E + 00	3.62519E + 01	5.76967E + 00	1.73320E − 01
4	2.12646E + 00	3.74206E + 01	5.95567E + 01	1.67907E − 01
5	2.13441E + 00	3.77010E + 01	6.00030E + 00	1.66658E − 01

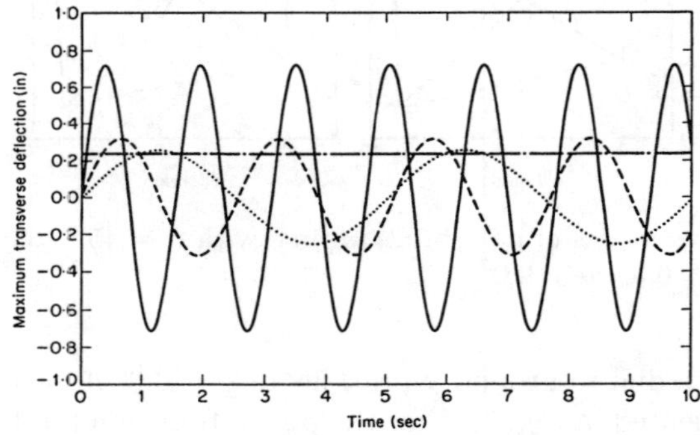

Fig. 7.9. Response spectra for forced vibration of a portal frame to a force 250 lb $K = 0.777$ —, $k = 0.9K$; - - - -, $k = 0.7\,K$; · · · · · ·, $k = 0.5\,K$; − · −, static

presented results provide a lower bound which may be used as a comparison for other numerical results obtained from approximate methods such as the finite element method.

Comparison between the first five frequencies of in-plane vibrations in Chap. 4 and out-of-plane vibrations for the portal frames with and without x-brace is made. The results are listed in Tables 7.9 and 7.10, respectively.

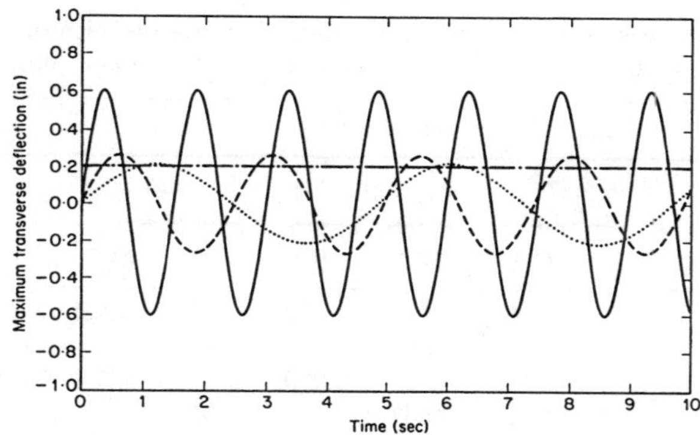

Fig. 7.10. Response spectra for forced vibration of an x-braced frame to a force 250 lb $K = 0.793$ —, $k = 0.9K$; - - - -, $k = 0.7K$; ⋯⋯, $k = 0.5K$; – · –, static

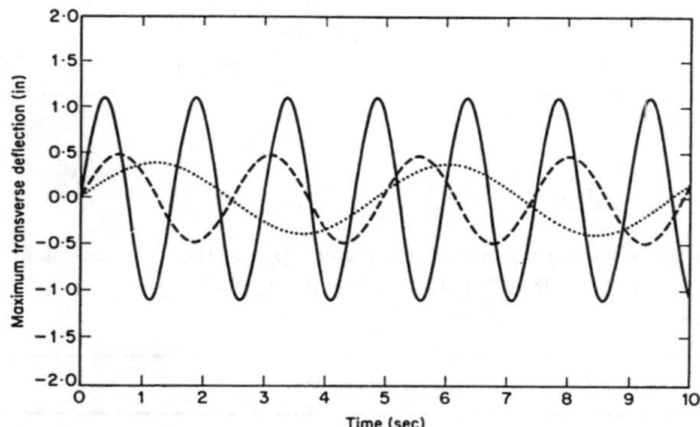

Fig. 7.11. Response spectra of a portal frame to horizontal ground vibrations with amplitude 0.36 in $K = 0.777$ —, $k = 0.9K$; - - - -, $k = 0.7K$; ⋯⋯, $k = 0.5K$

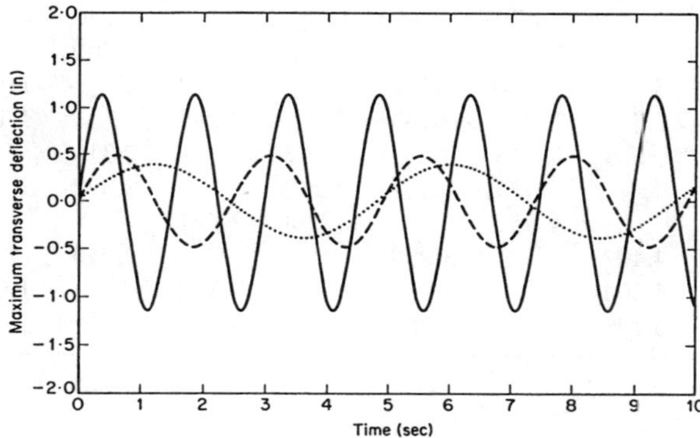

Fig. 7.12. Response spectra of x-braced frame to horizontal ground vibrations with amplitude 0.36 in $K = 0.793$ —, $k = 0.9K$; - - - -, $k = 0.7K$; ⋯⋯, $k = 0.5K$;

Table 7.9. Comparison of in-plane and out-of-plane frequencies of portal frames, all $A_i = 1.0$, $S_i = 1.0$ and $R = 50$, I-P = in-plane, O-P = out-of-plane

α (degrees)									
15		30		45		60		75	
I-P	O-P	I-P	O-P	I-P	O-P	I-P	O-P	I-P	O-P
	1.682		1.543		1.413		1.250		0.890
2.138		1.971		1.789		1.573		1.148	
	3.264		2.482		2.070		1.756	1.248	
	4.334		4.139	3.541		2.301			1.357
4.367		4.152			3.702		2.597		1.620
5.040			4.778		4.490	3.779		1.959	
	5.195	4.803		4.539			3.875		2.165
	7.332		6.243		4.687		4.233	2.719	
7.340			6.498		4.768	4.280			2.846
7.886		7.606		6.559		4.750		3.484	

Table 7.10. Comparison of in-plane and out-of-plane frequencies of x-braced frames, all $A_i = 1.0$, $S_i = 1.0$ and $R = 50$, I-P = in plane, O-P = out-of- plane

α (degrees)									
15		30		45		60		75	
I-P	O-P	I-P	O-P	I-P	O-P	I-P	O-P	I-P	O-P
	1.745		1.599		1.435		1.240		0.885
3.268			2.452		1.970		1.600		1.085
	3.314	3.670		3.022		0.129		1.112	
4.146		3.698		3.135		2.240		1.177	
	4.382		3.938		3.242		2.311		1.223
	4.537		4.009		3.260		2.322	1.241	
4.635			4.367		3.763	2.541			1.355
	4.643	4.407		3.983			2.649		1.511
4.659		4.504		4.269		3.590		1.892	
5.020		5.138		4.701		3.761		1.972	

It is shown that there are no regular patterns for the changes. However, the results clearly show that the first frequencies are associated with the out-of-plane vibrations.

Exercise

1. Determine the first four normal modes of out-of-plane vibrations of the frame for the problem defined in Chap. 5.

8

Buckling of Inclined Columns

The stability of a simple truss subjected to a single force, P, as shown in Fig. 8.1 was investigated by Mises [40] as a basic model to illustrate some of the buckling phenomena of structures. This model is known as the Mises truss. Buckling of the truss has also been studied by Von Karman and Kerr [54] and Huang and Vahidi [27]. In these studies, either the bending or the axial thrust was neglected.

In the present study, the effects of both bending and normal thrust are taken into consideration. The simply supported ends are generalized to include other possible end conditions. Thus, in general, this model represents a frame rather than a truss.

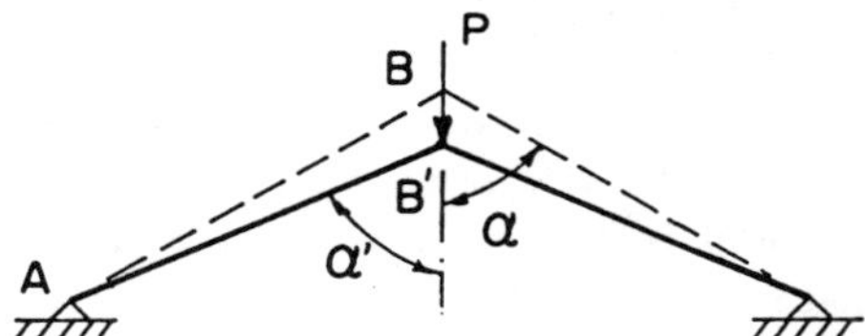

Fig. 8.1. A Mises truss

The frame is subjected to the vertical force, P, at the central joint as shown in Fig. 8.1. Because of symmetry, the structure may be analyzed by an inclined column as shown in Fig. 8.2(a). This column may also represent the inclined member of a three-bar frame with a rigid horizontal bar subjected to a single force at the center. This three-bar frame may simulate a meridional element of a conical frustum with rigid bulkheads undergoing axisymmetric deformation as shown in Fig. 8.3. The buckling behavior of conical shells will be investigated in Chap. 11.

The model shown in Fig. 8.2(a) is the same one used in Chap. 2 for vibrations, except that now the bar is subjected to the vertical force, F.

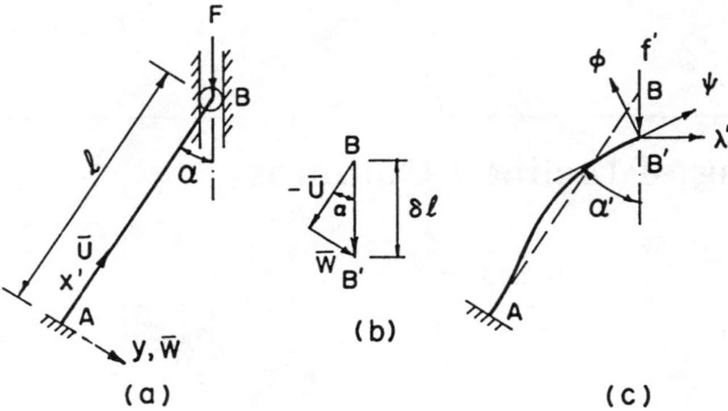

Fig. 8.2. An inclined bar

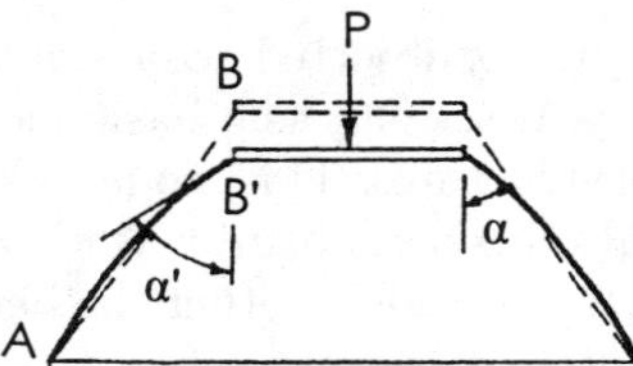

Fig. 8.3. A conical frustum

The inclined column possesses the following two unique features: (1) The top end, B, is deformable only in the vertical direction, and (2) the buckling strength is affected by the angle change from α to α' due to deformation as shown in these figures.

In the present study, by using the variation method incorporated with Lagrange's multiplier to account for the constraint condition as in Chap. 2, a set of nonlinear equations is derived. The exact solutions for the non-linear equations are obtained for inclined columns with various end conditions. The buckling behaviors of an inclined column with two ends simply supported are examined in detail, including the bifurcation from an axial to lateral bending configurations. The results are also compared with those obtained from eigenvalue method and other two conventional approaches.

8.1 Large Deflection Theory for Buckling of Inclined Columns

Consider the elastic and prismatic inclined column, AB, of the length l shown in Fig. 8.2(a) subjected to the vertical force, F. Let x and y be the axial and lateral coordinates with the origin located at the centroid of the cross-section at the lower end, A, as shown in Fig. 8.2(a); let $\bar{u}$ and $\bar{w}$ be the axial and lateral displacements, respectively; and let $\alpha(0 \leq \alpha < \pi/2)$ be the angle between the center line of the column and the vertical line. End A is fixed in space while the upper end, B, may have a displacement only in the vertical direction. Thus end B has a geometrical constraint which requires that the total displacement in the horizontal direction must vanish. Hence, as shown in Fig. 8.2(b)

$$\bar{u}\sin\alpha + \bar{w}\cos\alpha = 0 \quad \text{at} \quad x = l \tag{8.1}$$

Including the bending effect for large lateral deflection, the axial strain may be given as

$$\bar{e}_x = \bar{u}_{,x} + \frac{1}{2}\bar{w}_{,x}^2 - y\bar{w}_{,xx} \tag{8.2}$$

in which a subscript preceded by a comma represents the derivative with respect to the subscript. The potential energy or the functional [23] of this system, consisting of the strain energy accounting for $\bar{e}_x$, the work done by the force, F, and the constraint of (8.1), may be constructed.

Let A = cross-sectional area; E = Young's modulus; R = the slenderness ratio; and λ = the dimensionless Lagrange's multiplier. Physically, λ is the constraint force acting in the horizontal direction at the deformable end. Then the functional takes the following form:

$$V = AEl\left\{\frac{1}{2}\int_0^1\left[U_{,\eta} + \frac{1}{2}W_{,\eta}^2\right]^2 d\eta + \frac{1}{2}\frac{1}{R^2}\int_0^1 W_{,\eta\eta}^2 d\eta \right.$$
$$\left. + \left[f\left(U\cos\alpha - W\sin\alpha\right) - \lambda\left(U\sin\alpha + W\cos\alpha\right)\right]_{\eta=1}\right\} \tag{8.3}$$

in which

$$\eta = \frac{x}{l} \quad U = \frac{\bar{u}}{l} \quad W = \frac{\bar{w}}{l} \tag{8.4}$$

and

$$f = \frac{\mathrm{F}}{AE} \tag{8.5}$$

Let

$$\psi = U_{,\eta} + \frac{1}{2}W_{,\eta}^2 \tag{8.6}$$

be the axial strain or dimensionless axial force. The solution of the differential equation resulted from the vanish of the variation of functional V with respect to U yields

$$\psi = -\left(f\cos\alpha - \lambda\sin\alpha\right) \tag{8.7}$$

which indicates that the axial force in the column is a constant. Integrating (8.6) yields

$$U(\eta) = \psi\eta - \frac{1}{2}\int_0^\eta W_{,\eta}^2 d\eta \tag{8.8}$$

Vanishing of the variation V with respect to W leads to the differential equation for W

$$W_{,\eta\eta\eta\eta} + \beta^2 W_{,\eta\eta} = 0 \quad \text{for} \quad 0 \le \eta \le 1 \tag{8.9}$$

and the following boundary conditions:

$$\begin{array}{llll} W_{,\eta} = 0 & \text{or} \quad W_{,\eta\eta} = 0 & \text{at} \quad \eta = 0 & (8.10\text{a,b}) \\ W = 0 & \text{or} \quad W_{,\eta\eta\eta} + \beta^2 W_{,\eta} = 0 & \text{at} \quad \eta = 0 & (8.11\text{a,b}) \\ W_{,\eta} = 0 & \text{or} \quad W_{,\eta\eta} = 0 & \text{at} \quad \eta = 1 & (8.12\text{a,b}) \\ W = 0 & \text{or} \quad W_{,\eta\eta\eta} + \beta^2 W_{,\eta} = \phi R^2 & \text{at} \quad \eta = 1 & (8.13\text{a,b}) \end{array}$$

in which

$$\beta^2 = -\psi R^2 \quad \text{and} \tag{8.14}$$

$$\phi = -(f \sin\alpha + \lambda \cos\alpha) \tag{8.15}$$

The ϕ is the shear force in the column at top, B. Also, there is the constraint of (8.1) which now is

$$U \sin\alpha + W \cos\alpha = 0 \quad \text{at} \quad \eta = 1 \tag{8.16}$$

End A has no displacements and end B is deformable ($W \neq 0$). Therefore, the conditions of (8.11a and 8.13b) must be satisfied by all columns to be considered. The solution of (8.9) satisfying (1la and 13b) is

$$W(\eta) = B(\cos\beta\eta - 1) + C \sin\beta\eta - \frac{\phi}{\psi}\eta \tag{8.17}$$

By taking ψ as the independent parameter of the problem, the three constants, B, C, and ϕ in (8.17) may be determined by the two conditions, one each from (8.10 and 8.12) incorporated with (8.16).

Solving f and λ from (8.7 and 8.15), normalizing them with respect to the Euler buckling load [48] for the column at $\alpha = 0$, and then denoting the results by τ and γ, respectively, one has

$$\tau = \frac{fR^2}{\beta_0^2} = -\frac{R^2}{\beta_0^2}(\psi \cos\alpha + \phi \sin\alpha) \tag{8.18}$$

$$\gamma = \frac{\lambda R^2}{\beta_0^2} = \frac{R^2}{\beta_0^2}(\psi \sin\alpha - \phi \cos\alpha) \tag{8.19}$$

The coefficient, β_0, depends on the particular column under consideration. For $\alpha = 0$ and $\tau = 1$, (8.14 and 8.18) lead to $\beta = \beta_0$ (see later). For $\alpha = \pi/2$, the present formulations are still valid. However, they represent another class of problem – beam problems. Therefore, $\alpha = \pi/2$ is excluded from the present investigation.

The head shortening of the column shown in Fig. 8.2(b) is

$$\delta = -U \cos\alpha + W \sin\alpha \quad \text{at} \quad \eta = 1 \tag{8.20}$$

A curve of τ versus δ may be constructed. The maximum value of the curve τ, denoted by τ_1, is the buckling load for a given column.

In conventional analysis of conical shells, the constraint of (8.16) has been treated as a boundary condition replacing a condition equivalent to (8.13b) [6, 43]. For comparison, such conventional approach for inclined columns is given later along with the Mises approach.

8.2 Solutions

The solutions of the inclined column with various end conditions are presented in the following.

8.2.1 Inclined Columns with Two Ends Simply Supported

For simply supported columns, boundary conditions (8.10b and 8.12b), respectively, are

$$W_{,\eta\eta}(0) = 0 \quad \text{and} \quad W_{,\eta\eta}(1) = 0 \tag{8.21a,b}$$

Equation (8.21a) is satisfied when (8.17) becomes

$$W(\eta) = C\sin\beta\eta - \frac{\phi}{\psi}\eta \tag{8.22}$$

Equation (8.21b) may be fulfilled by this solution either $C = 0$ or $C \neq 0$, which requires

$$\beta = \pi \tag{8.23}$$

These two cases are considered separately.

When $C = 0$, (8.22) yields

$$W(\eta) = -\frac{\phi}{\psi}\eta \tag{8.24}$$

From (8.8), one has

$$U(\eta) = \left[\psi - \frac{1}{2}\left(\frac{\phi}{\psi}\right)^2\right]\eta \tag{8.25}$$

Equation (8.16) results into the following characteristic equation:

$$\phi = \psi\left[-\cot\alpha + \left(\cot^2\alpha + 2\psi\right)^{\frac{1}{2}}\right] \tag{8.26}$$

Equations (8.18 and 8.20) yield the following two equations, respectively, when (8.23, 8.24 and 8.25) are used

$$\tau = -\psi\left[\cos^2\alpha + 2\psi\sin^2\alpha\right]^{\frac{1}{2}}(R/\pi)^2 \tag{8.27}$$

$$\delta = \left[\cos\alpha - \left(\cos^2\alpha + 2\psi\sin^2\alpha\right)\right]^{\frac{1}{2}}\frac{1}{\sin\alpha} \tag{8.28}$$

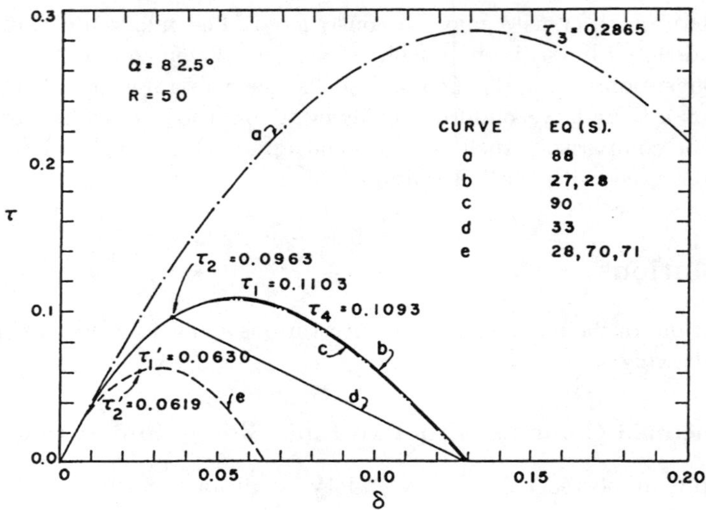

Fig. 8.4. Force vs. head-shortening curves for a column with simply supported ends[⊗]

In (8.27), $\beta = \pi$ was used. Equations (8.27 and 8.28) provide a τ versus δ curve as shown in Fig. 8.4.

The τ_1 determined from such a curve, in general, is higher than that denoted by τ_2 obtained from the solutions with $C \neq 0$. For instance, when $R = 100$ and $\alpha = 60°$, $\tau_1 = 32.4$ and $\tau_2 = 0.4985$. However, for large α and small R, these two buckling loads may be closer. A complete τ verses δ curve for $R = 50°$ and $\alpha = 82.5°$ is shown in Fig. 8.4 by curve b from which $\tau_1 = 0.1103$ and $\tau_2 = 0.0963$.

When $C \neq 0$, $\beta = \pi$. From (8.14), when ψ reaches

$$\psi_2 = -(\pi/R)^2 \tag{8.29}$$

a buckling configuration

$$W(\eta) = C \sin \pi\eta - (\tau - \cos\alpha) \frac{1}{\sin\alpha}\eta \tag{8.30}$$

is about to take place. In (8.30), (8.18) was used to replace ϕ by τ.

[⊗] In Fig. 8.4, for CURVE a, EQ 88 means EQ 8.88; so are for other curves and in other chapters.

The corresponding axial displacement, from (8.8), is

$$U(\eta) = -\left(\frac{\pi}{R}\right)^2 \eta - \frac{1}{4}\left[C^2\pi^2\left(\eta + \frac{1}{2\pi}\sin 2\pi\eta\right)\right.$$
$$\left. -4C(\tau - \cos\alpha)\frac{\sin\pi\eta}{\sin\alpha} + 2(\tau - \cos\alpha)^2 \frac{\eta}{\sin^2\alpha}\right] \tag{8.31}$$

By using (8.16), the amplitude, C, is determined as

$$C = \pm\sqrt{2}\left(\frac{1}{\pi\sin\alpha}\right)\left[\cos^2\alpha - 2\left(\frac{\pi}{R}\sin\alpha\right)^2 - \tau^2\right]^{\frac{1}{2}} \tag{8.32}$$

There is perhaps no way to determine the sign for C in (8.32). Thus unlike the other three types of inclined columns considered later, the solution of the present type is not unique. The buckling pattern is probably determined by the initial imperfections of the column.

Substituting (8.30 and 8.31) into (8.20) after rearrangement, one has

$$\tau = \cos\alpha - \delta\sin^2\alpha \tag{8.33}$$

The buckling load, τ_2, is determined by using (8.29) in (8.18 and 8.26). This gives

$$\tau_2 = \left[\cos^2\alpha - 2\left(\frac{\pi}{R}\sin\alpha\right)^2\right]^{\frac{1}{2}} \tag{8.34a}$$

Since $\tau_2 > 0$, the present buckling pattern is possible if

$$R > \sqrt{2}\pi\tan\alpha \tag{8.34b}$$

which is indicated in Fig. 8.5. The curve in Fig. 8.5 divides the R-α plane into two regions. Inclined columns falling in region I may have bifurcation type buckling; those in region II, buckling is due to direct thrust only.

When ψ increases to ψ_2 in (8.29), then τ and δ increase to τ_2 and δ_2 which are determined by (8.33). At $\tau = \tau_2$, the amplitude, C, is still zero. As δ increases further, (8.33) indicates that τ drops, and from (8.32), C starts to grow. Thus the bifurcation occurs continuously.

Equation (8.33) is shown in Fig. 8.4 by the straight line d for that column. Equation (8.34a) is presented in Fig. 8.6.

By taking the positive C, the elastic curves of (8.30 and 8.31) are depicted in Fig. 8.7 for a column of $\alpha = 30°$ and $R = 100$ at and after buckling. Fig. 8.7 indicates that after buckling as δ increases and τ drops slightly, the amplitude grows rather fast.

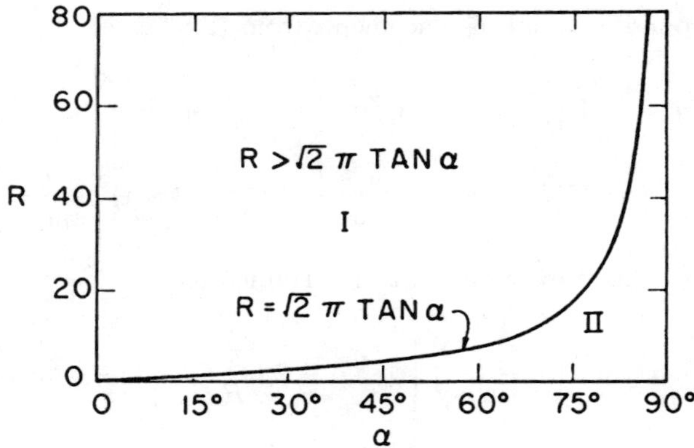

Fig. 8.5. Curve of equation (8.34b)

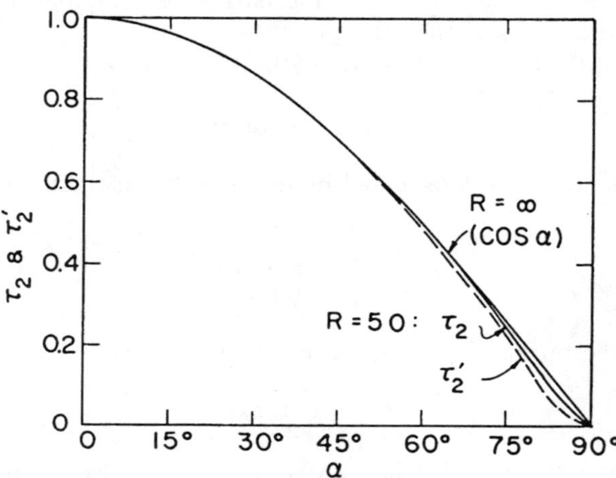

Fig. 8.6. Buckling loads of columns with simply supported ends

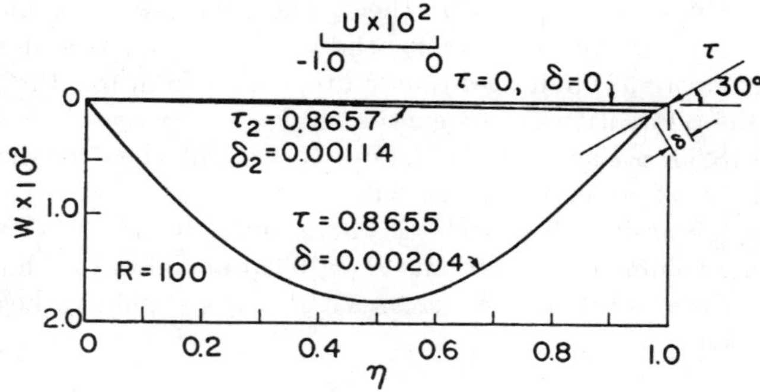

Fig. 8.7. Elastic curves of column with simply supported ends $\alpha = 30°$

8.2.2 Inclined Columns with Lower End Fixed and Upper End "Hinged"

A column as shown in Fig. 8.2(a) is considered. For such a column one requires

$$W_{,\eta}(0) = 0 \quad \text{and} \quad W_{,\eta\eta}(1) = 0 \tag{8.35a,b}$$

Then (8.17 and 8.8) lead to, respectively,

$$W(\eta) = \frac{\phi}{\beta\psi}\left[\sin\beta\eta - \beta\eta - \tan\beta\left(\cos\beta\eta - 1\right)\right] \tag{8.36}$$

$$U(\eta) = \psi\eta - \frac{1}{4\beta}\left(\frac{\phi}{\psi}\right)^2\left[\tan^2\beta\left(\beta\eta - \sin\beta\eta\cos\beta\eta\right) + 3\beta\eta - \tan\beta\left(3 + \cos 2\beta\eta - 4\cos\beta\eta\right) + \sin\beta\eta\left(\cos\beta\eta - 4\right)\right] \tag{8.37}$$

Equation (8.16) yields the following characteristic equation:

$$\phi = \frac{\psi}{2S}\left[-T\cot\alpha \pm \left(T^2\cot^2\alpha + 4S\psi\right)^{\frac{1}{2}}\right] \tag{8.38}$$

in which

$$S = \tfrac{1}{4}\left[3\left(1 - \tfrac{\tan\beta}{\beta}\right) + \tan^2\beta\right] \tag{8.39}$$

$$T = \tfrac{1}{\beta}\left(\beta - \tan\beta\right) \tag{8.40}$$

It has been found from numerical computation that when $T < 0$, the minus sign in front of the radical in (8.38) should be used; when $T > 0$, the positive sign should be taken. [When β is a complex number and the solutions are in hyperbolical functions, the sinusoidal functions in (8.36 and 8.37) shall be replaced by their corresponding hyperbolical functions.]

Equation (8.20) results in

$$\delta = \left[-\psi + S\left(\frac{\phi}{\psi}\right)^2\right]\cos\alpha - T\frac{\phi}{\psi}\sin\alpha \tag{8.41}$$

For the present case [48] from (8.40), when $T = 0$, then $\tan\beta = \beta$ yields

$$\beta_0 = 4.4934 \tag{8.42}$$

A set of τ versus δ curves is given in Fig. 8.8 for $R = 100$ and for various α angles. Curves for τ_1 versus α for $R = 50$, 100, and 200 are presented in Fig. 8.9.

Two deformed configurations before and at buckling are depicted in Fig. 8.10 for $R = 100$ and $\alpha = 30°$.

Note that in the present case and for the other two cases which follow, no bifurcations occurred so that the deformed elastic curves are unique.

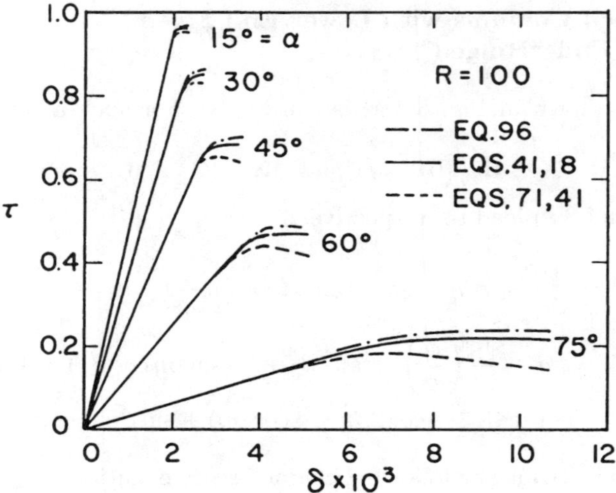

Fig. 8.8. Force vs. head-shortening curves of columns with upper end "hinged" and lower end fixed

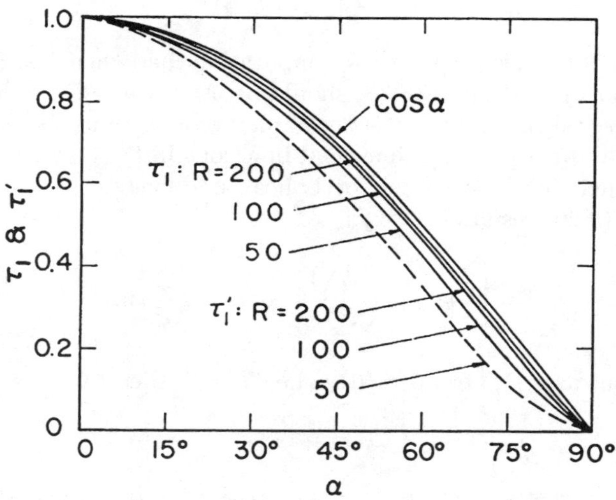

Fig. 8.9. Buckling loads of columns with lower end fixed and other end simply supported

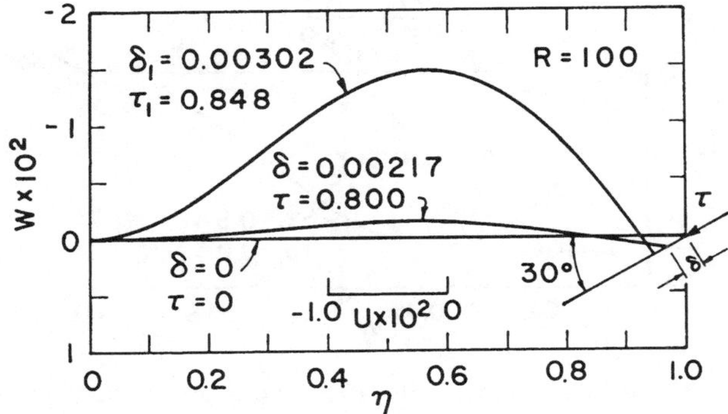

Fig. 8.10. Elastic curves of column with lower end fixed and other end simply supported $\alpha = 30°$

8.2.3 Inclined Columns with Lower End Hinged and Upper End Fixed in Rotation

For these columns, one requires

$$W_{,\eta\eta}(0) = 0 \quad \text{and} \quad W_{,\eta}(1) = 0 \tag{8.43a,b}$$

The lateral and axial displacements are

$$W(\eta) = \frac{\phi}{\beta\psi}\left(\frac{\sin\beta\eta}{\cos\beta} - \beta\eta\right) \tag{8.44}$$

and

$$U(\eta) = \psi\eta - \frac{1}{4}\frac{1}{\beta\cos^2\beta}\left(\frac{\phi}{\psi}\right)^2 \times \left[\beta\eta\left(1 + 2\cos^2\beta\right) - \left(4\cos\beta - \cos\beta\eta\right)\sin\beta\eta\right] \tag{8.45}$$

which yield the same values at $\eta = 1$ as (8.36 and 8.37) do. Thus these two types of columns share (8.38 to 8.42) as well as Figs. 8.8 and 8.9. However, the deformed configurations are different as shown in Fig. 8.11.

8.2.4 Inclined Columns with Lower End Fixed and Upper End Fixed in Rotation

For inclined columns with the lower end fixed and the upper end fixed in rotation but deformable in the vertical direction, (8.10 and 8.12) take the following forms:

$$W_{,\eta}(0) = 0 \quad \text{and} \quad W_{,\eta}(1) = 0 \tag{8.46a,b}$$

The resulted displacements are

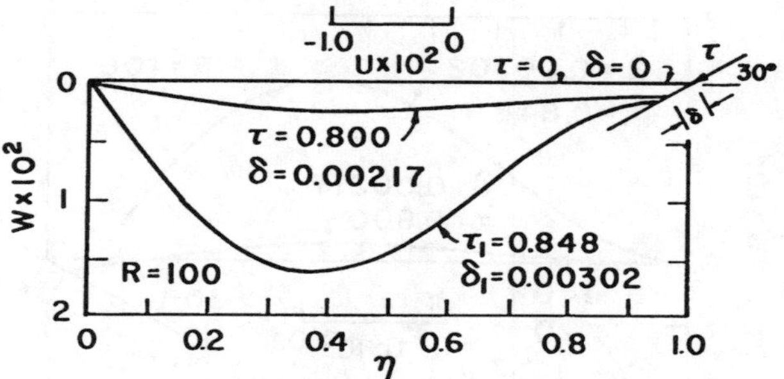

Fig. 8.11. Elastic curves of column with lower end simply supported and other end fixed in rotation $\alpha = 30°$

$$W(\eta) = \frac{\phi}{\beta\psi}\left[\sin\beta\eta - \beta\eta + (1-\cos\beta)\frac{1-\cos\beta\eta}{\sin\beta\eta}\right] \tag{8.47}$$

$$U(\eta) = \psi\eta - \frac{1}{8\beta}\left(\frac{\phi}{\psi}\right)^2 \{[(1-\cos\beta)(2\beta\eta - \sin 2\beta\eta)/\sin\beta - (6 - 8\cos\beta\eta - 2\cos 2\beta\ \eta)](1-\cos\beta)/\sin\beta + 6\beta\eta - 8\sin\beta\eta + \sin 2\beta\eta\} \tag{8.48}$$

Equations (8.38 to 8.41) are also applicable for the present case, except that

$$S = \frac{1}{4\beta}\frac{1}{\sin^2\beta}[(1-\cos\beta)(\beta - \beta\cos\beta - 3\sin\beta) - 3\sin\beta(\cos 2\beta - \cos\beta - \beta\sin\beta)] \tag{8.49}$$

$$T = \frac{1}{\beta\sin\beta}[\beta\sin\beta - 2(1-\cos\beta)] \tag{8.50}$$

For $T = 0$, one has

$$\beta_0 = 2.860\pi \tag{8.51}$$

which is also shown in [48]. The related numerical results for this column are presented in Figs. 8.12, 8.13, and 8.14.

8.3 Eigenvalue Method

In addition to the nonlinear force-displacement method in the determination of the buckling loads of a column, the linearized eigenvalue method is also often used in structural mechanics. For comparison, the eigenvalues of the inclined columns are computed in the following.

Let a set of perturbed displacements, u and w, be superposed on the displacements, U and W, respectively. At this disturbed state, a functional

$$\bar{V} = V + v \tag{8.52}$$

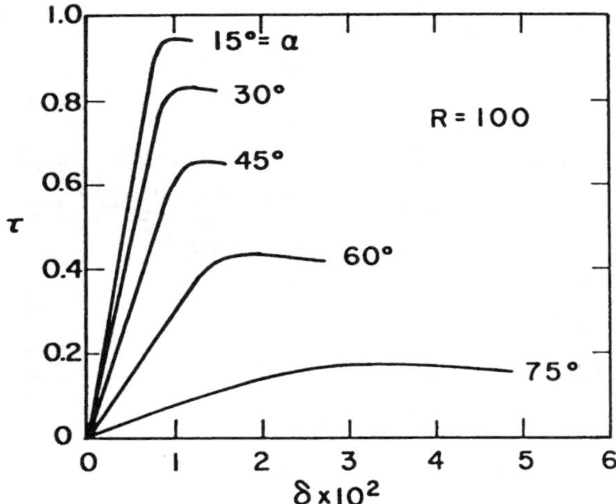

Fig. 8.12. Forced vs. head-shortening curves of columns with lower end fixed and other end fixed in rotation

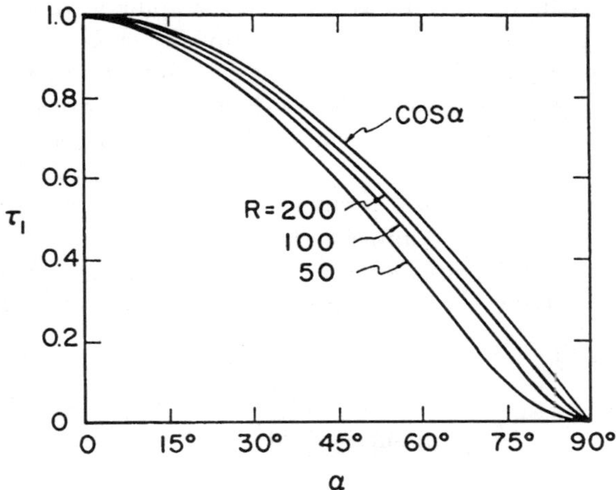

Fig. 8.13. Buckling loads of columns with lower end fixed and other end fixed in rotation

may be constructed by substituting $U + u$ and $W + w$ for U and W, respectively, into (8.3). Subtracting V as given by (8.3) from $\bar{V}$ results in

$$v = AEl \left\{ \frac{1}{2} \int_0^1 \left[u,_{\eta}^2 - \psi w,_{\eta}^2 \right] d\eta + \frac{1}{2R^2} \int_0^1 w,_{\eta\eta}^2 \, d\eta - \rho \left[u(1) \sin\alpha + w(1) \cos\alpha \right] \right\} \tag{8.53}$$

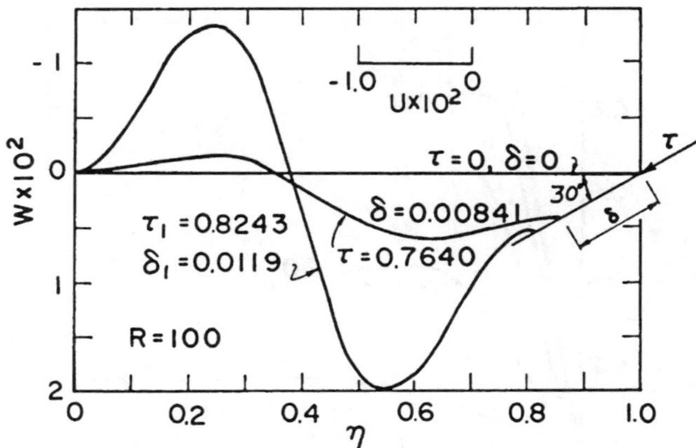

Fig. 8.14. Elastic curves of column with lower end fixed and other end fixed in rotation $\alpha = 30°$

in which, because of the small values of u and w, their higher order terms have been neglected including a term, $W,_{\eta}\ w,_{\eta}$, for simplicity. [The W, thus $W,_{\eta}$ is small for small α, but not for large α. This simplicity probably accounts for the bigger differences in the values of β in Table 8.1, and for larger values of α, the differences get bigger. This also agrees with the observations made in [48] for shallow arches. Thus for large values of α, the linearized perturbation method cannot be applied.]

In (8.53), the ρ is another Lagrange's multiplier representing the change of the constraint force at the upper end.

Analogous to the previous approach, the following two sets of equations are obtained by letting the variation of v with respect to u and w vanish, respectively:

$$u,_{\eta\eta} = 0 \qquad \text{for} \quad 0 \le \eta \le 1 \tag{8.54}$$

$$u,_{\eta} = \rho \sin\alpha \qquad \text{at} \quad \eta = 1 \tag{8.55}$$

$$u = 0 \qquad \text{at} \quad \eta = 0 \tag{8.56}$$

$$\text{and} \quad \mathrm{w},_{\eta\eta\eta\eta} + \beta^2 \mathrm{w},_{\eta\eta} = 0 \qquad \text{for} \quad 0 \le \eta \le 1 \tag{8.57}$$

$$w = 0 \quad \text{or} \quad w,_{\eta\eta\eta} + \beta^2 w,_{\eta} = 0 \quad \text{at} \quad \eta = 0 \tag{8.58a,b}$$

$$w,_{\eta\eta} = 0 \quad \text{or} \quad w,_{\eta} = 0 \quad \text{at} \quad \eta = 0 \tag{8.59a,b}$$

$$w = 0 \quad \text{or} \quad w,_{\eta\eta\eta} + \beta^2 w,_{\eta} + \rho R^2 \cos\alpha = 0 \quad \text{at} \quad \eta = 1 \tag{8.60a,b}$$

$$w,_{\eta\eta} = 0 \quad \text{or} \quad w,_{\eta} = 0 \quad \text{at} \quad \eta = 1 \tag{8.61a,b}$$

in which β is defined in (8.14). From (8.54 to 8.56)

$$u = \rho \sin \alpha\eta \tag{8.62}$$

The solution of (8.57) is

Table 8.1. Eigenvalues for inclined columns[⊗]

		β for Columns in Solutions 2 and 3		β for Columns in Solution 4	
α, in degrees (1)	R (2)	τ-δ curve (3)	Eq. (66) (4)	τ-δ curve (5)	Eq. (68) (6)
0	–	–	4.4934	–	8.9860
15	200	4.4855	4.4934	8.9576	8.9868
	100	4.4810	4.4934	8.9246	8.9859
	50	4.4651	4.4933	8.8622	8.9865
30	200	4.4766	4.4934	8.9196	8.9865
	100	4.4632	4.4933	8.8516	8.9856
	50	4.4320	4.4928	8.7090	8.9820
45	200	4.4676	4.4933	8.8678	8.9859
	100	4.4396	4.4930	8.7466	8.9832
	50	4.3853	4.4916	8.4860	8.9720
60	200	4.4452	4.4931	8.7817	8.9841
	100	4.4000	4.4921	8.5574	8.9758
	50	4.3011	4.4879	8.0622	8.9396
75	200	4.3909	4.4918	8.5440	8.9740
	100	4.2895	4.4870	8.0000	8.9312
	50	4.0249	4.4656	6.4614	8.6691

$$w(x) = a\cos\beta\eta + b\sin\beta\eta + c\,\beta\eta + d \tag{8.63}$$

The five constants, a, b, c, d, and ρ, are to be determined by the four boundary conditions of (8.58 to 8.61) and one constraint

$$u(1)\sin\alpha + w(1)\cos\alpha = 0 \tag{8.64}$$

Eigenvalues β are obtained by the condition of non-trivial solutions of the five constants.

For the inclined columns with two ends simply supported, one eigenvalue is $\beta = \pi$ as given by (8.23), and the other, following the present approach, is

$$\beta = R\cot\alpha \tag{8.65}$$

which may be applied only when α is close to $\pi/2$.

For the inclined columns described in cases 8.2.2 and 8.2.3, both result in the following eigenvalue equation:

$$(\beta - \tan\beta)/\beta^3 = \tan^2\alpha/R^2 \tag{8.66}$$

[⊗] In Table 8.1, Eq. 66 and Eq. 68 mean Eq. (8.66) and Eq. (8.68); so are for equations in other tables and other chapters.

which reduces to

$$\beta = \tan\beta \quad \beta = 4.493 = \beta_0 \tag{8.67}$$

for $\alpha = 0$ or $R = \infty$ as given in [48], and also by (8.42).

For the inclined columns with both ends fixed, the eigenvalue equation is

$$\frac{\beta^2}{R^2}\tan^2\alpha + 2\frac{1-\cos\beta}{\beta\sin\beta} - 1 = 0 \tag{8.68}$$

which for $\alpha = 0$ becomes

$$2(\cos\beta - 1) + \beta\sin\beta = 0 \tag{8.69}$$

as given in [48]. The eigenvalue of (8.69) is 2.860π as shown in (8.51).

Table 8.1 lists the roots of (8.66 and 8.68) along with β values obtained from the force displacement method.

8.4 Effect of Angle Change

As a basic principle for stability problems, the equations for buckling are supposedly derived at a deformed state. For instance, when a vertical column is subjected to an axial force, the column has axial deformation but remains straight until buckling takes place. But for an inclined column, once the external force is applied on, the lateral displacement will be induced by the axial displacement because of the constraint condition. In turn, the inclined angle will be changed. This will reduce the buckling strength of the column. The reduction is investigated now.

For a given axial force, ψ, of a given inclined column, the solutions can be obtained including the corresponding shear force, ϕ. Now the question is how much of the vertical force, f, required to produce so much ψ and ϕ? If the original state is used as the reference, as it was done earlier, (8.18) provides the answer. If the deformed state is used as the reference, then the angle, α, at the top end becomes α' as shown in Figs. 8.1 and 8.2(c):

$$\alpha' = \alpha + W_{,\eta} \qquad \text{at} \quad \eta = 1 \tag{8.70}$$

It should not be confused that the prime in the last equation, and hereafter, merely makes a distinguish for the same symbols used in the previous sections.

The corresponding values of τ and γ denoted by τ' and γ' may be obtained as follows:

$$\tau' = \frac{f'R^2}{\beta_0^2} = -\frac{R^2}{\beta_0^2}(\psi\cos\alpha' + \phi\sin\alpha') \tag{8.71}$$

$$\gamma' = \frac{\lambda' R^2}{\beta_0^2} = \frac{R^2}{\beta_0^2}(\psi\sin\alpha' - \phi\cos\alpha') \tag{8.72}$$

Thus for improvement, instead of τ versus δ curves, the τ' versus δ curves should be used in the determination of the buckling loads. Such τ' versus δ curves for cases 8.2.1 and 8.2.2 are shown by dashed curves in Figs. 8.4 and 8.8.

For the other two types of columns, because of the vanishing of the 2nd term in (8.70), there is no angle change.

For column with two simply supported ends, the buckling load, τ_2', determined from (8.24, 8.26, 8.29 and 8.71), is

$$\tau_2' = \sin\alpha' \left\{ \cot\alpha' - \cot\alpha + \left[\cot^2\alpha - 2\left(\frac{\pi}{R}\right)^2 \right]^{\frac{1}{2}} \right\} \tag{8.73}$$

in which

$$\alpha' = \alpha + \cot\alpha - \left[\cot^2\alpha - 2\left(\frac{\pi}{R}\right)^2 \right]^{\frac{1}{2}} \tag{8.74}$$

Equation (8.73) is given in Fig. 8.6 for $R = 50$.

For the column of Fig. 8.4 with $\alpha = 82.5°$, $\tau_2' = 0.0619$ is the least buckling load ever computed.

8.5 Comparison with Conventional Approach

Following the approach used in the analysis of conical shells [43], the axial force, ψ, is not statically indeterminate as it is in the previous formulations. Instead, the axial force, now denoted by ψ' is given as

$$\psi' = -\frac{f}{\cos\alpha} = -\frac{\tau\beta_0^2}{R^2\cos\alpha} \tag{8.75}$$

The reaction, λ', is supposedly cancelled out with the other component, i.e.

$$\lambda' = f\tan\alpha \tag{8.76}$$

which is not involved in the rest of the formulations.

Substitution of (8.75) into (8.8), the axial displacement is

$$U' = -\frac{f}{\cos\alpha}\eta - \frac{1}{2}\int_0^{\eta} W'^2_{,\eta}\, d\eta \tag{8.77}$$

The lateral displacement W' is governed by the differential equation

$$W'_{,\eta\eta\eta\eta} + \Omega^2 W'_{,\eta\eta} = 0 \tag{8.78}$$

and boundary conditions:

$$W' = 0 \qquad \text{at} \quad \eta = 0 \tag{8.79}$$

$$W'_{,\eta} = 0 \quad \text{or} \quad W'_{,\eta\eta} = 0 \qquad \text{at} \quad \eta = 0 \tag{8.80a,b}$$

$$W'_{,\eta} = 0 \quad \text{or} \quad W'_{,\eta\eta} = 0 \qquad \text{at} \quad \eta = 1 \tag{8.81a,b}$$

$$U'\sin\alpha + W'\cos\alpha = 0 \qquad \text{at} \quad \eta = 1 \tag{8.82}$$

In (8.78)

$$\Omega^2 = \frac{fR^2}{\cos\alpha} = \frac{\tau\beta_0^2}{\cos\alpha} \tag{8.83}$$

A comparison of (8.77 to 8.82) with (8.8 to 8.16) reveals that when $\alpha = 0$, both sets of equations reduce to the same equations for the Euler column [48]. But when $\alpha = \pi/2$, because of (8.75), this set of equations is not defined, while (8.8 to 8.16) reduce to a set of equations for a class of beams as mentioned earlier.

The solution of (8.78) is

$$W' = H\cos\Omega\eta + G\sin\Omega\eta + I\Omega\eta + J \tag{8.84}$$

Two numerical examples are given: (1) the Mises truss and (2) the column shown in Fig. 8.2(a).

8.5.1 Mises Truss

With the satisfaction of (8.79, 8.80b, 8.81b and 8.82), the solutions for the Mises truss are

$$U' = -\left(\tfrac{\tau\pi^2}{R^2\cos\alpha} + \tfrac{1}{2}N^2\right)\eta \tag{8.85}$$

$$W' = N\eta \tag{8.86}$$

in which

$$N = \cot\alpha - \left(\cot^2\alpha - \frac{2\tau\pi^2}{R^2\cos\alpha}\right)^{\frac{1}{2}} \tag{8.87}$$

Substitution of these results into (8.20) yields

$$\tau = \left(\frac{R}{\pi}\right)^2 \delta\cos\alpha\left(\cos\alpha - \frac{1}{2}\delta\sin^2\alpha\right) \tag{8.88}$$

which is presented by curve a in Fig. 8.4 for that column. The buckling load determined by this curve is denoted by τ_3 which is much higher than the other values computed.

Mises [40] also used (8.75) to determine the axial force. The deformed angle α' is used, i.e.

$$\psi'' = -\frac{f}{\cos\alpha'} \tag{8.89}$$

Therefore, the axial force, ψ'', is statically indeterminate. But in this formulation, no second term in (8.77) is included. The result, in the present notation, is

$$\tau = \left(\frac{R}{\pi}\right)^2 (\cos\alpha - \delta)\left[\left(1 - 2\delta\cos\alpha + \delta^2\right)^{-\frac{1}{2}} - 1\right] \tag{8.90}$$

which is depicted in Fig. 8.4 by curve c and the buckling load is denoted by τ_4 for this column. This curve, in this particular case, almost coincides with curve b.

8.5.2 For an Inclined Column

For the inclined column shown in Fig. 8.2(a), (8.84 and 8.77) satisfying (8.79, 8.80a, 8.81b, and 8.82) are, respectively,

$$W'(\eta) = G(\sin\Omega\eta - \tan\Omega\cos\Omega\eta - \Omega\eta + \tan\Omega) \tag{8.91}$$

$$\begin{aligned} U'(\eta) = &-\left(\frac{\beta_0}{R}\right)^2 \frac{\tau}{\cos\alpha}\eta - \frac{G^2\Omega^2}{\psi}\left[\left(3+\tan^2\Omega\right)\eta\right. \\ &+\frac{1}{2\Omega}\left(1-\tan^2\Omega\right)\sin 2\Omega\eta \\ &\left.+\frac{\psi}{\Omega}\sin\Omega\eta + \frac{\tan\Omega}{\Omega}\left(\cos\Omega\eta - 1\right)\left(\cos\Omega\eta + 3\right)\right] \end{aligned} \tag{8.92}$$

in which

$$G = \frac{1}{2S}\left\{T\cot^2\alpha \pm \left[T^2\cot^2\alpha - 4S\frac{\tau}{\cos\alpha}\frac{\beta_0}{R}\right]^{\frac{1}{2}}\right\} \tag{8.93}$$

$$S = \frac{1}{\psi}\Omega\left[3\left(\Omega - \tan\Omega\right) + \Omega\tan^2\Omega + 8\sin\Omega\right] \tag{8.94}$$

$$T = \tan\Omega - \Omega \tag{8.95}$$

For $T < 0$, the plus sign in front of the radical in (8.93) is taken; for $T > 0$, the minus sign is used. Equation (8.20) yields

$$\delta = \left[\frac{\tau}{\cos\alpha}\left(\frac{\beta_o}{R}\right)^2 + SG^2\right]\cos\alpha + TG\sin\alpha \tag{8.96}$$

which is shown in Fig. 8.8 by the dash-dot curves. The buckling loads determined from these curves are also consistently higher.

This comprehensive study presents the exact solutions of a class of large deflection problems. The buckling loads determined from these solutions are lower than those obtained by the eigenvalue method. The latter, nevertheless, provides the simple equation for the eigenvalue, β, which in turn yields the buckling value of ψ for a given column. In fact, the eigenvalues are practically applicable for columns with small angle α and large slenderness ratio R. Table 1 indicates that for the larger angle, the differences get bigger, and if the effect of angle change is taken into consideration, the story will be quite different for both ends simply supported inclined columns.

Exercises

1. For an inclined column, $\alpha = 45°$ and $R = 100$, with the bottom end simply supported, the top end may freely rotate but can be deformed in the vertical direction. The column is subjected to a vertical force at the top end. Determine the buckling loads τ_1, τ_2, and τ_1' as shown in Fig. 8.4.
2. Verify the statements that $\tau_1 = 32.4$ and $\tau_2 = 0.4985$ for a column with simply supported ends for $R = 100$ and $\alpha = 60°$.

9

Inclined Girders with End Constraint Subjected to Normal Loads

A two-bar truss subjected to lateral normal distributed loads as shown in Fig. 9.1 represents a double leaf miter-type gate used in navigation locks as shown in Fig. 9.2. The gate is subjected to hydraulic static pressure whose intensity depends on the level of depth. The gate is supported by horizontal girders. Both ends of the girder are hinged and fixed in space, and the center joint is also hinged but deformable along the central line. Due to symmetry, the girder may be analyzed by an inclined girder subjected to the lateral distributed normal load as shown in Fig. 9.3(a).

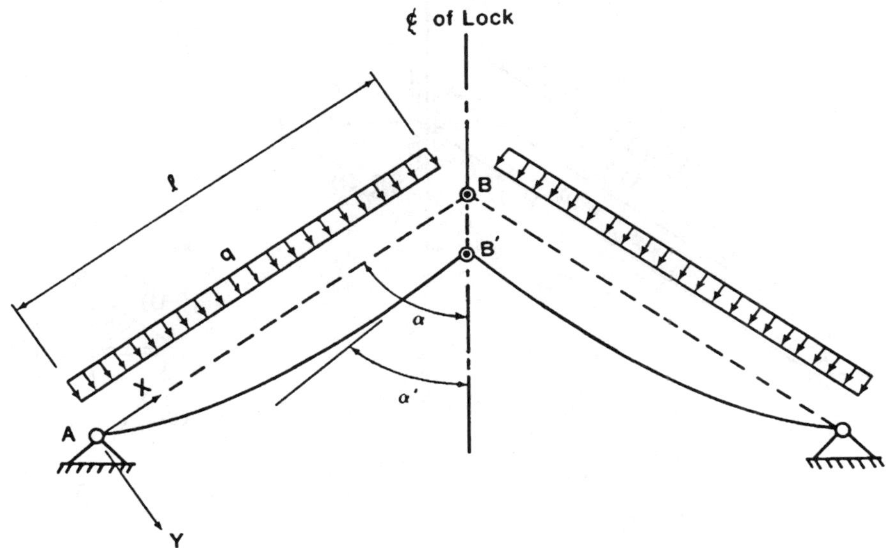

Fig. 9.1. Deformed mitered gate girders

Fig. 9.2. A double leaf miter-type gate (Courtesy of Corps of Engineers)

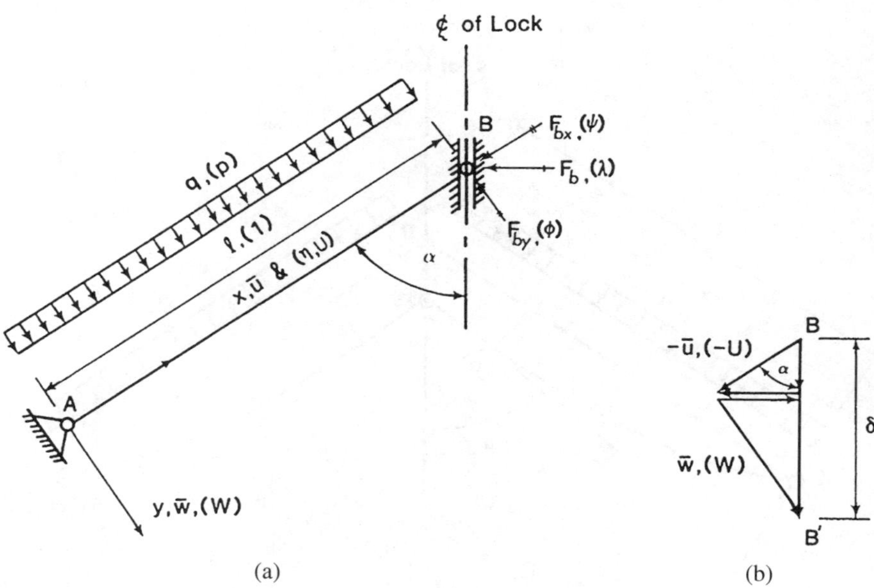

Fig. 9.3. An inclined girder with end constraint

The current design of the gate is based on the Corps of Engineers' Engineering Manual [17]. In this design, each girder is treated as a three-hinged arch. The girder is considered as a beam-column subjected to combined axial thrust and lateral uniformly distributed load.

The above design did not consider the effect of the end deformation from B to B' and the effect of angle change from α to α' as shown in Fig. 9.1. The present study is to examine these effects.

It has been found in the last chapter that the reduction of strength of the columns due to angle change, however, is relatively small for most ranges of α; only when α is close to 90°, the effect is appreciable. Thus, for simplicity in the present study, the angle change will not be considered in the general formulations. This effect will be examined later for girders which have angles near to 90°.

9.1 Formulation and Solutions

Consider a horizontally framed miter-type gate in its mitered position, with the gate supported by two mitered girders at the closed position. The hydrostatic pressure is uniformly distributed and is symmetric about the center line of the lock. Because of symmetry, a typical girder as shown in Fig. 9.3(a) is studied. This model is similar to the inclined bar studied in the last chapter of buckling of inclined columns, except that the vertical force now is replaced by the uniformly distributed lateral load. Therefore, the bending effect including axial thrust due to the arch action to this inclined girder is investigated in the present chapter.

The girder is assumed to be elastic, with length l, section modulus S, and cross-sectional area A. The distributed load of intensity, q, per unit length acts in the plane passing through the horizontal principal axis of the cross section.

In Fig. 9.3, $\bar{u}$ and $\bar{w}$ are the displacements in the axial, x, and lateral, y, directions, respectively. At end B, there is a geometrical constraint which requires the end can be deformed only along the center line of the lock, i.e., the horizontal displacement which is normal to the center line of the lock at end B must vanish. Thus from Fig. 9.3(b)

$$\bar{u} \sin \alpha + \bar{w} \cos \alpha = 0 \qquad \text{at } x = 1 \tag{9.1}$$

in which $\alpha(0 < \alpha < \pi/2)$ is the angle between the center line of the girder and the axis of the lock.

Including the bending effect due to the axial and large lateral displacement, the axial strain is

$$\bar{e}_x = \bar{u}_{,x} + \frac{1}{2}\bar{w}_{,x}^2 - y\bar{w}_{,xx} \tag{9.2}$$

in which a subscript preceded by a comma represents the derivative with respect to the subscript.

Introducing

$$\eta = \frac{x}{l} \quad U = \frac{\bar{u}}{l} \quad W = \frac{\bar{w}}{l} \tag{9.3}$$

and

$$p = q/EI \tag{9.4}$$

then a functional of this system consisting of the potential energy, the strain energy, external potential energy due to load q, and the constraint of (9.1), may be formulated as

$$V = AEl\left[\frac{1}{2}\int_0^1 \left(U_{,\eta} + \frac{1}{2}W_{,\eta}^2\right)^2 d\eta + \frac{1}{2R^2}\int_0^1 W_{,\eta\eta}^2\, d\eta \right.$$
$$\left. - \frac{pl^2}{A}\int_0^1 W d\eta - \lambda\left(U\sin\alpha + W\cos\alpha\right)_{\eta=1}\right] \tag{9.5}$$

in which E, R, and λ are Young's modulus, slenderness ratio, and Lagrange's multiplier, respectively. Physically, λ is the reaction at end B, or F_b in Fig. 9.3(a), normal to the axis of the lock.

By the variation of V with respect to U, one obtains

$$\psi = \lambda \sin\alpha \tag{9.6}$$

where

$$\psi = U_{,\eta} + \frac{1}{2}W_{,\eta}^2 \tag{9.7}$$

which is the axial strain or dimensionless axial force, F_{bx}, as shown in Fig. 9.3(a). Equation (9.6) indicates that the axial force in the girder is constant along its length. Integration of (9.7) yields

$$U(\eta) = \psi\eta - \frac{1}{2}\int_0^\eta W_{,\eta}^2 d\eta \tag{9.8}$$

The vanishing of the variation of V with respect to W leads to the differential equation for W:

$$W_{,\eta\eta\eta\eta} + \beta^2 W_{,\eta\eta} = \frac{pl^2R^2}{A} \quad 0 \le \eta \le 1 \tag{9.9a}$$

and the following boundary conditions:

$$W = 0 \quad \text{and} \quad W_{,\eta\eta} = 0 \quad \text{at } \eta = 0 \tag{9.9b-d}$$
$$W_{,\eta\eta} = 0 \quad \text{and} \quad W_{,\eta\eta\eta} + \beta^2 W_{,\eta} = \phi R^2 \quad \text{at } \eta = 1 \tag{9.9e-g}$$

in which

$$\beta^2 = -\psi R^2 \tag{9.10}$$
$$\phi = -\lambda\cos\alpha \tag{9.11}$$

The ϕ is the dimensionless shearing force, F_{by}, as shown in Fig. 9.3(a). The constraint of (9.1) now is

$$U \sin\alpha + W \cos\alpha = 0 \quad \text{at } \eta = 1 \tag{9.12}$$

The solution of (9.9a) accounting for the boundary conditions of (9.9b to 9.9e) is

$$W(\eta) = \frac{Pl^2}{\beta^2 \phi A}\left(1 - \cos\beta\eta + H \sin\beta\eta\right) + \frac{G}{\psi}\eta - \frac{Pl^2}{2\psi A}\eta^2 \tag{9.13}$$

in which

$$H = \cot\beta - \csc\beta \tag{9.14}$$

$$G = \tfrac{Pl^2}{A} - \phi \tag{9.15}$$

Substituting (9.13) into (9.8) and integrating it result in

$$\begin{aligned} U(\eta) = \psi\eta - \frac{1}{2}\Bigg[& \frac{5P^2l^4}{2\beta^3\psi^2A^2}H + \frac{2Pl^2}{\beta^2\psi^2A}G + \frac{G^2}{\psi^2}\eta + \frac{P^2l^4}{2\beta^2\psi^2A^2}\left(H^2+1\right)\eta \\ & - \frac{Pl^2}{\psi^2 A}G\eta^2 + \frac{P^2l^4}{3\psi^2A^2}\eta^3 + \frac{2Pl^2}{\beta^2\psi^2A}\left(HG - \frac{Pl^2}{\beta A}\right)\sin\beta\eta \\ & - \frac{2Pl^2}{\beta^2\psi^2A}\left(G + \frac{Pl^2}{\beta A}H\right)\cos\beta\eta + \frac{P^2l^4}{4\beta^3\psi^2A^2}\left(H^2-1\right)\sin 2\beta\eta \\ & - \frac{P^2l^4}{\beta^2\psi^2A^2}\left(\frac{H}{2\beta}\cos 2\beta\eta + 2H\eta\sin\beta\eta - 2\eta\cos\beta\eta\right)\Bigg] \end{aligned} \tag{9.16}$$

The λ may be eliminated from (9.6 and 9.11). Then

$$\phi = -\psi\cot\alpha \tag{9.17}$$

Substitution of (9.13 and 9.16) into (9.12) yields a characteristic equation for ψ and P. Let P_y be the maximum dimensionless load, such that

$$P_y = \frac{8\sigma_y S}{El^3} \tag{9.18}$$

in which σ_y is the yielding stress. Normalizing P with respect to P_y, the characteristic equation may be presented in the following form:

$$\frac{P}{P_y} = \pm\frac{1}{P_y}\sqrt{\frac{b}{a}} \tag{9.19}$$

in which

$$\begin{aligned} a = \frac{l^4}{2\beta^2\psi^2A^2\sin^2\beta}\left(1-\cos\beta\right)\Bigg[& \left(1 - \frac{5}{\beta}\sin\beta\right) \\ & + \left(2 + \frac{\beta^2}{3}\right)\left(1+\cos\beta\right)\Bigg] \end{aligned} \tag{9.20}$$

$$b = \frac{1}{2}\cot^2\alpha + \psi \tag{9.21}$$

The "a" defined by (9.20) is an even function of ψ and is positive. Thus, in order to have a real value of P in (9.19), "b" defined in (9.21) must be positive. Therefore

$$\psi \geq -\frac{1}{2}\cot^2\alpha \tag{9.22}$$

which limits the range of ψ.

Two characteristic curves of P/P_y versus ψ from (9.19) for $\alpha = 70°$ and $88°$ are presented by dotted lines in Figs. 9.4(a,b), respectively, for a girder $l = 741.6$ in, $A = 50.94\,\text{in}^2$, $I = 22,305.33\,\text{in}^4$, and $S = 796.62\,\text{in}^3$. The corresponding deflection curves may be obtained from (9.13 and 9.16). Two typical elastic curves are shown in Figs. 9.5(a,b) for $\alpha = 70°$, $\psi = -0.2256 \times 10^3$ and $\alpha = 88°$, $\psi = -0.1805 \times 10^3$, respectively.

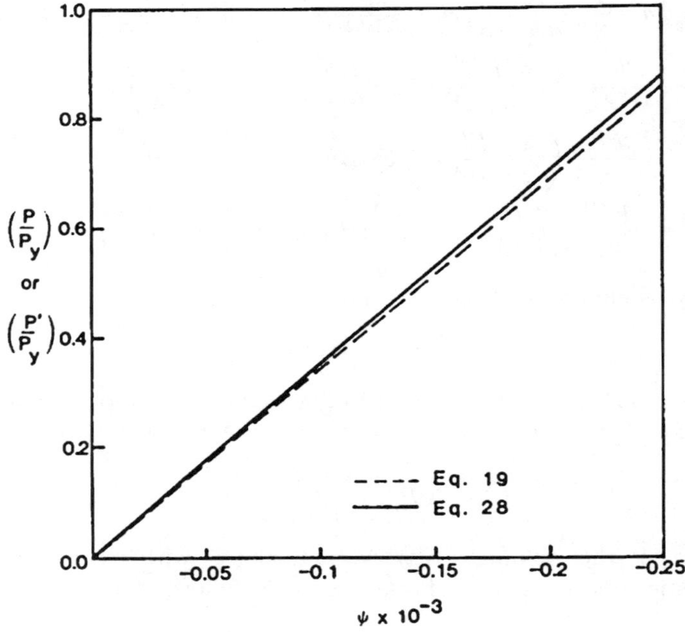

Fig. 9.4a. Characteristic curves of lateral load vs. axial strain $\alpha = 70°$

The complete curve shown in Fig. 9.4b accounts for both the positive and negative signs in (9.19). The negative branch has no physical meaning to this problem. The positive curve indicates that the axial thrust, ψ, is proportional to the lateral load initially. However, less lateral load is required to maintain the buckled configuration after the buckling load, P_{cr}/P_y, is reached.

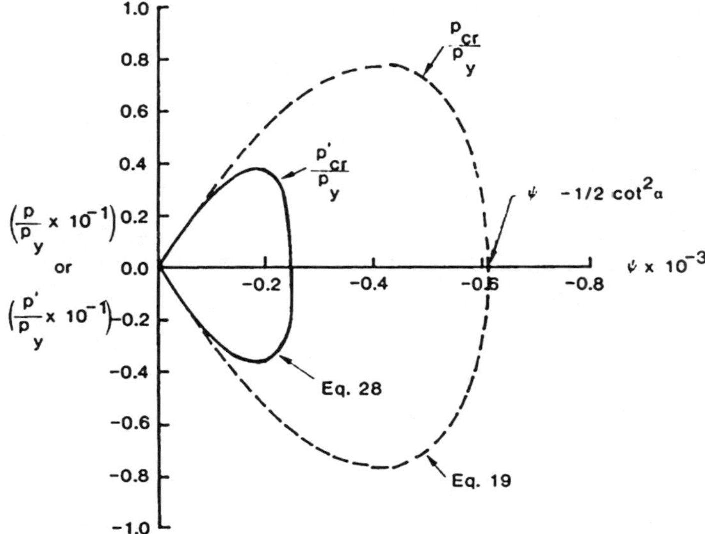

Fig. 9.4b. Characteristic curves of lateral load vs. axial strain $\alpha = 88°$

Let

$$\delta_e = \frac{\pi^2 \sec \alpha}{R^2} \tag{9.23}$$

be the deflection of a hinged column under the Euler buckling load. As shown in Fig. 9.3(b), the deflection of end B along the center line of the lock normalized with respect to δ_e is

$$\frac{\delta}{\delta_e} = \frac{1}{\delta_e}(-U \cos \alpha + W \sin \alpha) \quad \text{at } \eta = 1 \tag{9.24}$$

The load, P/P_y, versus head-shortening, δ/δ_e, curves are presented in Fig. 9.6a for the range of α from 40° to 86° by dotted lines; each curve is up to the limit where the yielding stress is developed. In Fig. 9.6b, a complete curve for $\alpha = 88°$ is depicted by the dotted line.

The maximum compressive stress, accounting for the axial thrust and moment and then normalized with respect to σ_y, is

$$\frac{\sigma_{\max}}{\sigma_y} = \frac{1}{\sigma_y}\left[-\frac{EIl}{AS}\frac{p}{\psi}(\cos \beta\eta - H \sin \beta\eta - 1) - \psi E\right] \tag{9.25}$$

Two sets of curves of P/P_y versus $\sigma_{\max}/\sigma_y$ are depicted in Figs. 9.7(a,b) for $\alpha = 40°$ to 86° and 88°, respectively,

9.2 Effect of Angle Change

In buckling analysis, a deformed state should be used as a reference. For the mitered gate girder as shown in Fig. 9.1, the initial angle, α, at end B would change to α' due to bending and axial thrust. It has been stated earlier that this angle change in α has little effect for small angles but when α is close to $\pi/2$, it is significant.

To study its effect, let the angle change from α to α' as shown in Fig. 9.1. When the effect of angle change is taken into consideration, a prime will be added to the corresponding symbols in what follows. Let

$$\alpha' = \alpha + W_{,\eta} \quad \text{at } \eta = 1 \tag{9.26}$$

From (9.13)

$$\alpha' = \alpha + \frac{Pl^2}{\psi A}\left[\frac{1}{\beta}\left(\sin\beta\eta + H\cos\beta\eta\right) - \eta\right] + \frac{G}{\psi} \tag{9.27}$$

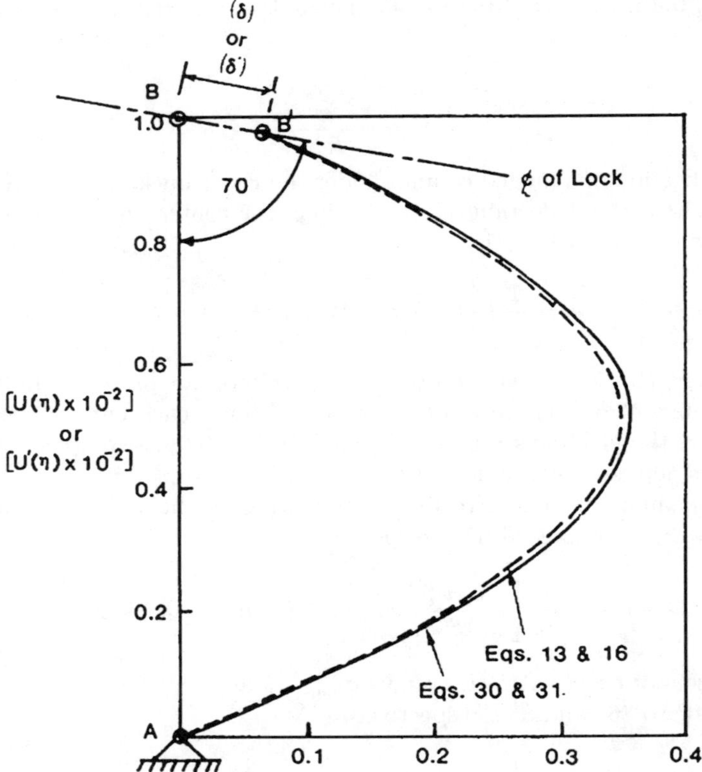

Fig. 9.5a. Elastic curves $\alpha = 70°$

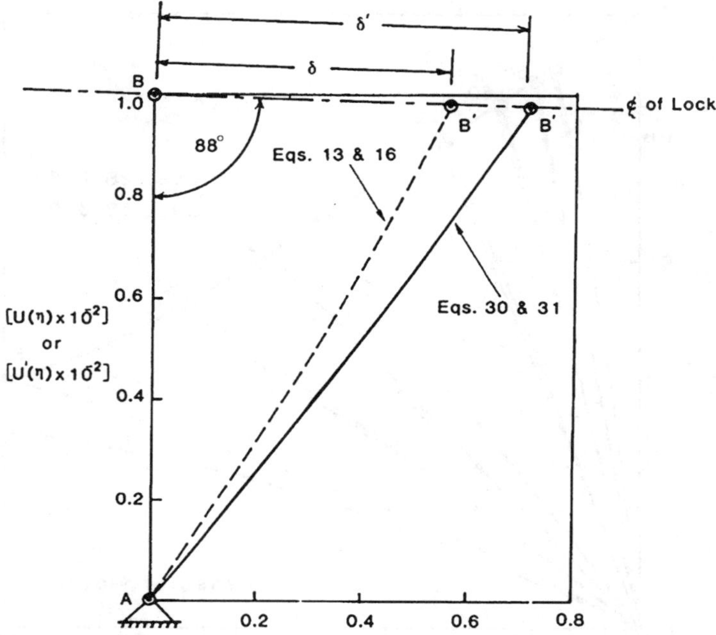

Fig. 9.5b. Elastic curves $\alpha = 88°$

Thus, (9.19) becomes

$$\frac{P'}{P_y} = \pm \frac{1}{P_y} \sqrt{\frac{b'}{a}} \tag{9.28}$$

in which

$$b' = \frac{1}{2} \cot^2 \alpha' + \psi \geq 0 \tag{9.29}$$

The displacements of (9.13 and 9.16) become, respectively,

$$W'(\eta) = \frac{G'}{\psi}\eta + \frac{P'l^2}{\beta^2 \psi A}\left(1 - \frac{1}{2}\beta^2\eta^2 - \cos\beta\eta + H\sin\beta\eta\right) \tag{9.30}$$

and

$$\begin{aligned} U'(\eta) = \psi\eta - \frac{1}{2}\Bigg[\frac{G'^2}{\psi^2}\eta + \frac{P'^2 l^4}{\beta^3\psi^2 A^2}\Bigg(&\frac{5}{2}H + \frac{1}{2}H^2\beta\eta + \frac{1}{2}\beta\eta + \frac{1}{3}\beta^3\eta^3 \\ &-2\sin\beta\eta - 2H\cos\beta\eta + \frac{1}{4}H^2\sin 2\beta\eta - \frac{1}{4}\sin 2\beta\eta \\ &-\frac{1}{2}H\cos 2\beta\eta - 2H\beta\eta\sin\beta\eta + 2\beta\eta\cos\beta\eta\Bigg) \\ &+ \frac{P'l^2}{\beta^2\psi^2 A}\left(2G' - G'\beta^2\eta^2 + 2HG'\sin\beta\eta - 2G'\cos\beta\eta\right)\Bigg] \end{aligned} \tag{9.31}$$

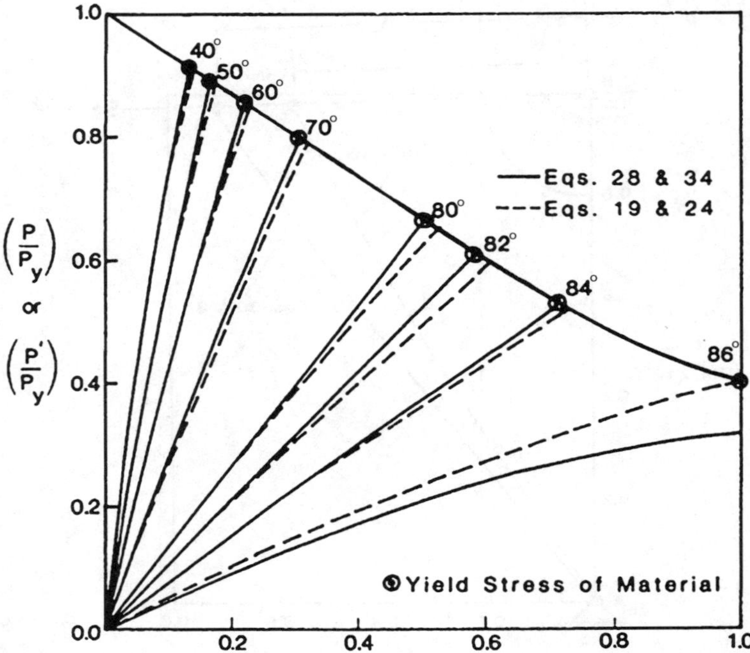

Fig. 9.6a. Characteristic curves of lateral load vs. head-shortening $\alpha = 40^\circ$ to 86°

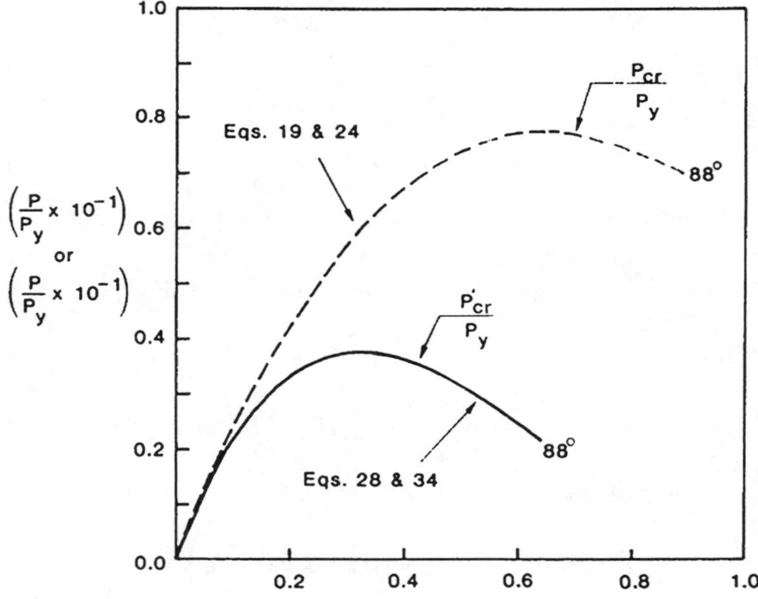

Fig. 9.6b. Characteristic curves of lateral load vs. head-shortening $\alpha = 88^\circ$

in which

$$G' = \frac{P'l^2}{A} - \phi' \tag{9.32}$$

where

$$\phi' = -\psi \cot \alpha' \tag{9.33}$$

The head-shortening becomes

$$\frac{\delta'}{\delta_e} = \frac{1}{\delta_e}\left(-U' \cos \alpha' + W' \sin \alpha'\right) \quad \text{at } \eta = 1 \tag{9.34}$$

The critical moment and maximum compressive stress now take the following forms, respectively,

$$M_c' = -\frac{EIl}{A}\frac{P'}{\psi}\left(\cos \beta\eta - H \sin \beta\eta - 1\right) \tag{9.35}$$

$$\frac{\sigma'_{\max}}{\sigma_y} = \frac{1}{\sigma_y}\left[\frac{EIl}{AS}\frac{P'}{\psi}\left(\cos \beta\eta - H \sin \beta\eta - 1\right) - \psi E\right] \tag{9.36}$$

The effects are shown in Figs. 9.4 to 9.7 by solid curves. These curves indicate that the angle change stiffens the structure for small angles, while for a large angle ($\alpha > 84°$), it softens the structure and buckling may occur before the maximum stress reaches the yielding value as shown for $\alpha = 88°$ in Figs. 9.7b and 9.8b.

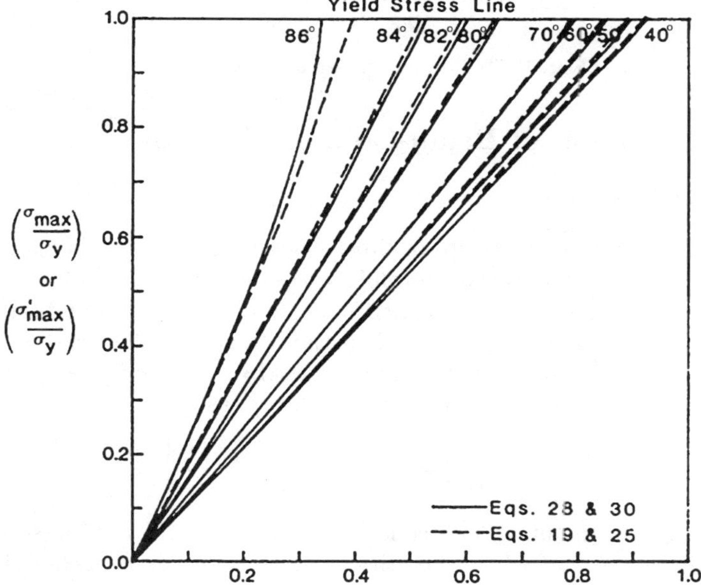

Fig. 9.7a. Maximum stress curves $\alpha = 40°$ to $86°$

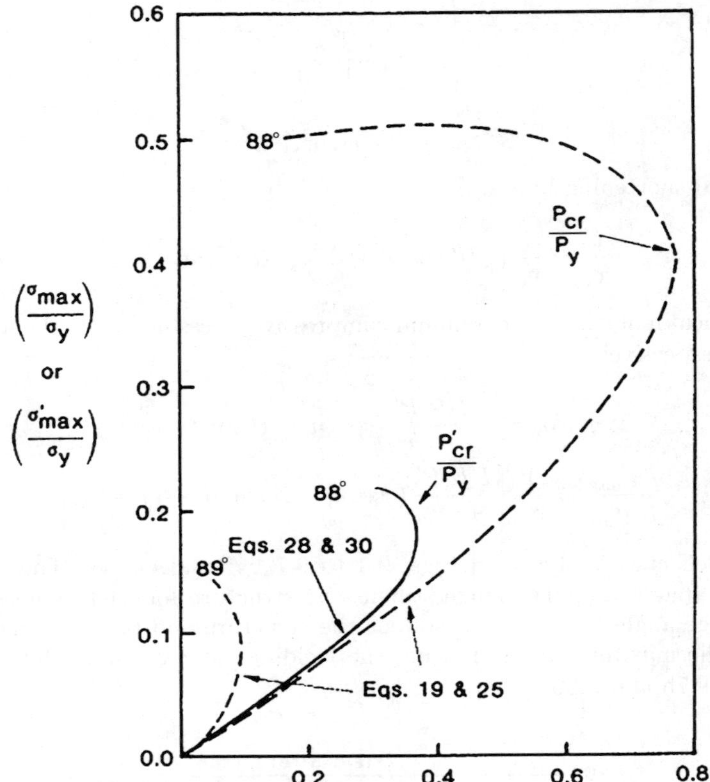

Fig. 9.7b. Maximum stress curves $\alpha = 88°$

9.3 Comparison of Beam-Column Solutions

9.3.1 AISC Formula

The AISC (American Institute of Steel Construction) formula suggested by Chen [10] is applicable to the allowable stress design to estimate the ultimate strength of the members subjected to the combined axial compression and bending. The axial stress permitted in the absence of the bending moment is controlled by the yielding stress, and the AISC interaction formula can be modified after normalization with respect to σ_y, as

$$\frac{\sigma_{\max}}{\sigma_y} = \frac{1}{\sigma_y}\left[\frac{F_{bx}}{A} + \frac{A\sigma_e}{(A\sigma_e - F_{bx})}\frac{M_c}{S}\right] \leq 1 \tag{9.37}$$

in which $\sigma_{\max}$ is the maximum compressive stress at the critical section; F_{bx} is the axial force; and M_c is the critical moment at the center of the girder determined by the equilibrium method. The equilibrium is based on the assumption that the girder is a simply supported beam-column.

9.3.2 Cheong's Beam-Column Formula

Cheong [11] suggested a similar equation which was derived for the beam-column with respect to the yielding stress of material:

$$\frac{\sigma_{\max}}{\sigma_y} = \frac{1}{\sigma_y}\left[\frac{F_{bx}}{A} + \frac{q}{K^2 S}\frac{1}{\cos(Kl/2)} - 1\right] \tag{9.38}$$

in which

$$K^2 = \frac{F_{bx}}{EI} \tag{9.39}$$

Results of (9.37 and 9.38) are compared with those of the present investigation in Fig. 9.8.

From the present study, it may be concluded that since the angles of the existing lock gate structures are around 71°, no modification is needed for the present AISC formulas.

The present study provides exact solutions for comparison.

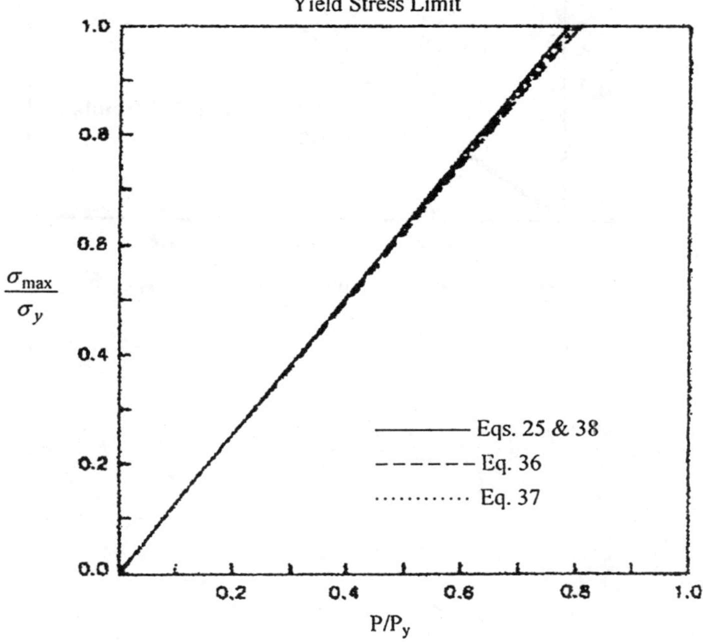

Fig. 9.8a. Comparison of maximum stress $\alpha = 70°$

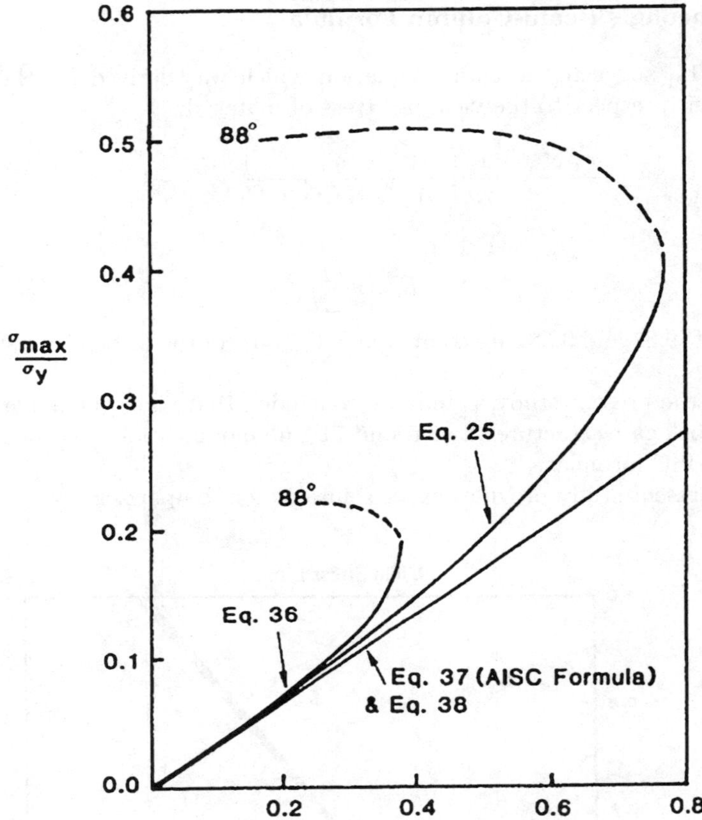

Fig. 9.8b. Comparison of maximum stresses $\alpha = 88°$

Exercise

1. Verify (9.13 and 9.19).

10

A Boundary-Layer Solution Verified by Buckling of a Rectangular Plate

The so called edge-zone or boundary-layer solutions are widely used in large deflection problems of thin elastic plates [20,35,36] and shells [31,35], but little experimental work has been seen in the literature to verify these approximate solutions. The boundary-layer solutions for the buckling of a rectangular plate are shown in [36]. An experiment was performed for this simple problem. The agreement between the analytical and experimental results are excellent.

Since this boundary-layer technique will be employed in the next chapter for the study of buckling of thin conical shells, both the analytic solutions and the experimental results of the buckling of the plate are presented in this chapter.

10.1 Basic Equations for Buckling of a Plate and Boundary-Layer Solution

The two Karman's equations for a dimensionless lateral deflection, W, and Airy's stress function, F, of an elastic plate undergoing large deflection may be given in the following form [48]:

$$\gamma^2 \nabla\nabla W - Q\left(F, W\right) = 0 \tag{10.1a}$$

$$\alpha^2 \nabla\nabla F + Q\left(W, W\right)/2 = 0 \tag{10.1b}$$

in which

$$\gamma^2 = \frac{1}{12}\left(1-\nu^2\right) \qquad \alpha = \frac{h}{A} \tag{10.2a,b}$$

$$Q\left(F, W\right) = F_{,xx}W_{,yy} + F_{,yy}W_{,xx} - 2F_{,xy}W_{,xy} \tag{10.2c}$$

where ν = Poisson's ratio, h = thickness of the plate, A = amplitude of deflection, and ∇ = Laplacian operator in Cartesian xy-coordinate as shown in Fig. 10.1; subscripts preceded by a comma represents derivatives with respect to the subscripts.

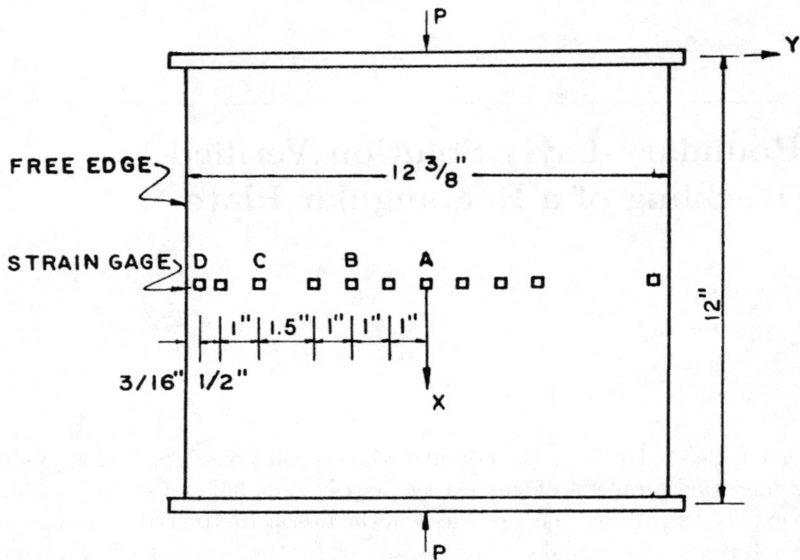

Fig. 10.1. The Plate ($1'' = 2.54\,\mathrm{cm}$)

The Airy's stress function, F, satisfies the equilibrium equations of plane stresses of the plate and is defined for the plane stresses in the following way:

$$N_{xx} = h^3 EF_{,yy} \quad N_{yy} = h^3 EF_{,xx} \quad N_{xy} = -h^3 EF_{,xy} \tag{10.3a-c}$$

where N_{xx}, N_{yy}, and N_{xy} are in plane normal and shearing stresses in the directions defined by the subscripts.

An $a \times 2b$ (12 in $\times$ 12(3/8) in) (30.48 cm $\times$ 31.4325 cm) plate as shown in Fig. 10.1 is considered. The plate has the top and bottom edges simply supported and the two side edges free. A pair of the single force, P, is applied on the opposite sides ($x = 0, a$) with two rigid rods which are so constructed that the rods can rotate freely.

The plate has the following boundary conditions. For the simply supported edges:

$$W = 0 \qquad \text{at } x = 0 \text{ and } a \tag{10.4a}$$

$$M_{xx} = -EAh^3\gamma^2 \left(W_{,xx} + \nu W_{,yy}\right) = 0 \qquad \text{at } x = 0 \text{ and } a \tag{10.4b}$$

and for the free edges:

$$M_{yy} = -EAh^3\gamma^2 \left(W_{,yy} + \nu W_{,xx}\right) = 0 \quad \text{at } y = \pm b \tag{10.5a}$$

$$Q_y = -EAh^3\gamma^2 \left(W_{,xx} + W_{,yy}\right)_{,y} = 0 \quad \text{at } y = \pm b \tag{10.5b}$$

where M_{xx}, M_{yy} are the normal moments, and Q_y is the shearing force in the directions defined by the subscripts.

Functions

$$W^0 = \sin\frac{\pi x}{a} \quad F^0 = -\gamma^2\pi^2\frac{y^2}{2a^2} \tag{10.6a,b}$$

satisfy (10.1) and all the required boundary conditions except conditions (10.5), since

$$W^0_{,xx}(x, \pm b) \neq 0 \tag{10.7}$$

Now let

$$W = W^0(x) + \alpha W^1(x,\eta) + \dots \tag{10.8a}$$
$$F = F^0(y) + f^0(x,\eta) + \alpha f^1(x,\eta) + \dots \tag{10.8b}$$

in which

$$\eta = (b - y)\,\alpha^{-\frac{1}{2}} \tag{10.9}$$

is a boundary-layer coordinate measured from $y = +b$. Substituting series (10.8) into (10.1) results in a set of equations associated with different orders of α. The set with the lowest order of α has the following equations:

$$\gamma^2 W^1_{,\eta\eta\eta\eta} - W^0_{,xx} f^0_{,\eta\eta} = 0 \tag{10.10a}$$
$$W^0_{,xx} W^1_{,\eta\eta} + f^0_{,\eta\eta\eta\eta} = 0 \tag{10.10b}$$

With the following boundary conditions at the free edge ($\eta = 0$):

$$W^1_{,\eta\eta} + \nu W^0_{,xx} = 0 \quad W^1_{,\eta\eta\eta} = 0 \tag{10.11a,b}$$
$$f^0_{,xx} = 0 \quad f^0_{,x\eta} = 0 \tag{10.12a,b}$$

The solutions satisfying (10.10 to 10.12) and the required conditions are

$$W^1(x,\eta) = \nu\gamma e^{-\lambda\eta}(\cos\lambda\eta - \sin\lambda\eta) \tag{10.13a}$$
$$f^0(x,\eta) = -\nu\gamma^2 e^{-\lambda\eta}(\cos\lambda\eta + \sin\lambda\eta) \tag{10.13b}$$

in which

$$\lambda^2 = \frac{1}{2\gamma}|W^0_{,xx}| \tag{10.14}$$

The normal stresses at the central section ($x = a/2$), from (10.3b), are

$$N_{xx} = h^3 E\left(F^0 + f^0\right)_{,yy}$$
$$= -h^3 E(\gamma\pi/a)^2 + AE(\pi\alpha)^2\nu\gamma e^{-\lambda\eta}(\cos\lambda\eta - \sin\lambda\eta) \tag{10.15a}$$
$$N_{yy} = h^3 E f^0_{,xx} = -E(\pi\alpha)^3\nu\gamma(\gamma/2)^{1/2}\eta e^{-\lambda\eta}\sin\lambda\eta \tag{10.15b}$$

Thus N_{yy} vanishes at the free edge ($\eta = 0$) and also at the center of the plate ($\eta = \infty$); furthermore, N_{yy} is of higher order of α in comparison with N_{xx}. Therefore, its effect is negligible in computation with normal strain e_{xx}. Thus

$$e_{xx} = (N_{xx} - \nu N_{yy})/Eh \approx N_{xx}/Eh \tag{10.16}$$

which is to be compared with the experimental results.

10.2 Comparison of Analytic Solutions with Experimental Results

A plate of bare (unclad) aluminum sheet 2024-T3 of $h = 1/16\,\text{in}$ (1.587 mm) was used in the experiment. The modulus of elasticity, $E = 11.11 \times 10^6\,\text{psi}$ (76.60 Pa) and $\nu = 0.331$.

The experiment was performed by an Instron universal testing machine. The applied load was maintained at an almost constant value, a little lower than Euler buckling load of $P_e = 232\,\text{lb}$ (1023N) as shown in Fig. 10.2. An increasing amplitude was developed after the buckling by the increasing head-shortening. The strains at strain gages A, B, C, and D of Fig. 10.1 were taken. The results of e_{xx} after coverting to N_{xx} normalized with the average stress

$$N_{xx}^0 = P/(hb) \tag{10.17}$$

are depicted in Fig. 10.3 at different load stages. The broken lines in both Figs. 10.3 and 10.4 are merely drawn to fit the experimental data.

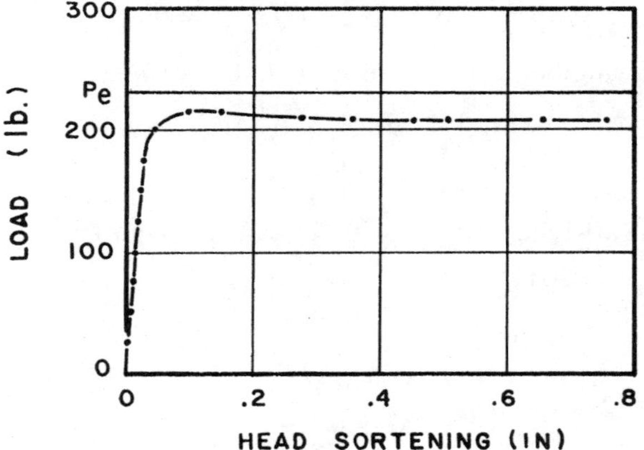

Fig. 10.2. Load vs. head-shortening

The increased tensile strain at gage D was compared with the analytical prediction; the agreement may be termed remarkable in the range tested. A comparison of the analytical solutions with the experimental results of the stress distribution at the central section of the plate at the largest value ever tested is presented in Fig. 10.4.

The solution given in (10.13a) is the additional deflection due to the anticlastic curvature, which is now confined along the two free edges with the maximum value at $\eta = 0$. For $\nu = 0.331$, the maximum value is

$$hW^1 = h\nu\gamma = 0.0902h \tag{10.18}$$

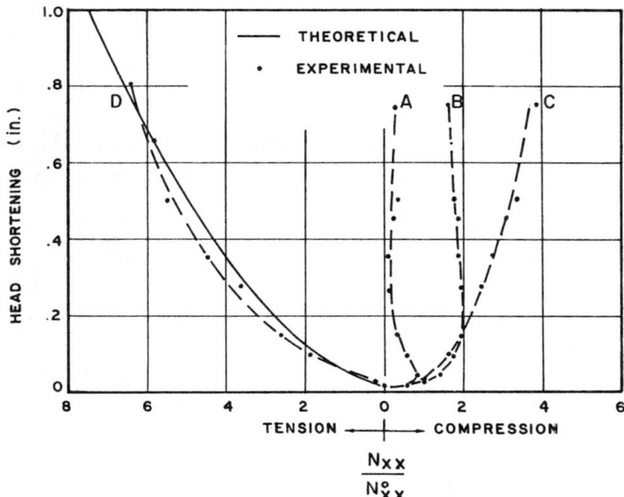

Fig. 10.3. Normal strains at strain gages A, B, C and D and from the boundary layer solution and experimental results

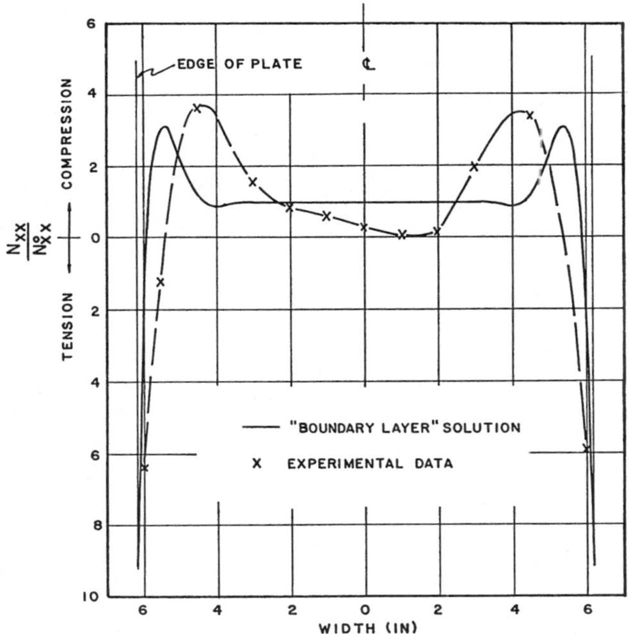

Fig. 10.4. Normal stresses at central section

Exercises

1. Show that functions (10.6a,b) satisfy (10.1a,b) and boundary conditions, $W = 0$ and $W_{,xx} = 0$, at $x = 0$.
2. Show that solutions (10.13) satisfy (10.10) and boundary conditions (10.11).
3. Explain that the plate is under compression but the maximum stresses at both free edges are in tension.
4. Try to obtain the boundary-layer solutions for $y = -b$.

11

Buckling of Conical Shells under a Single Axial Load

The buckling of conical shells subjected to external loads is a classic problem in shell structures. Since the theoretical predications could not agree with the experimental results, a number of research works have been performed [44]. In the present study, an improvement is made by following the two suggestions in Chaps. 2 and 8: (1) treating the smaller end constraint properly and (2) taking the effect of the angle change into consideration. It is shown that with these improvements, the theoretical predictions are closer to the experimental results.

The solution consists of two parts: membrane and bending. For the bending solution, the boundary-layer technique demonstrated in the last chapter is used.

11.1 Formulation of Large Deflection Theory for Buckling of Conical Shells

Consider the conical shell shown in Fig. 11.1(a). The truncated shell is subjected to the axial load, P. The cone has rigid bulkheads at top and bottom. Having the bottom or larger end fixed in space, the upper end, because of the axisymmetric deformation, can be deformed only in the vertical direction. This leads to the constraint

$$U \sin\alpha + W \cos\alpha = 0 \quad \text{at} \quad s = L \tag{11.1}$$

where U and W are the displacements in the meridional and normal directions of the middle surface of the cone, respectively, as indicated in Fig. 11.1(a). The α is the central angle; s is the meridional coordinate with the origin at the apex; and L is the distance from the apex to the smaller end as shown in Fig. 11.1(a).

For the axisymmetric case, the strain energy of the shell composed of membrane and bending deformations is

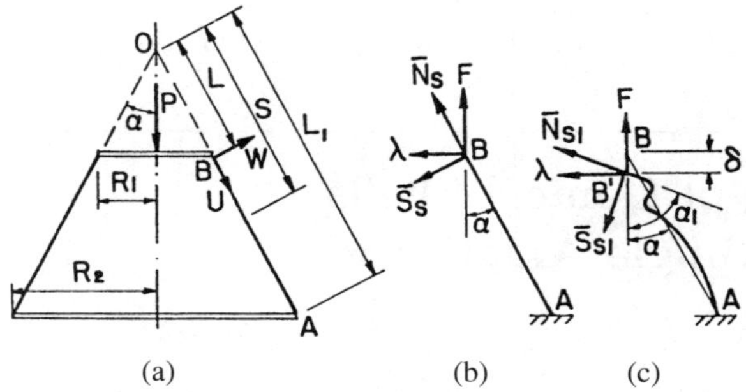

Fig. 11.1. A conical shell subjected to an axial load

$$V_s = \pi \sin\alpha \int_L^{L_1} (N_s e_s + N_\theta e_\theta + M_s K_s + M_\theta K_\theta) s ds \tag{11.2}$$

in which N, e, M, and K are the stress resultant, strain, moment resultant, and curvature of the middle surface of the shell, respectively, in the s and θ directions as indicated by the subscripts. The angle $\theta\,(0 < \theta \leq 2\pi)$ is the coordinate in the circumferential direction.

The potential energy of the external force, P, is

$$V_w = 2\pi L \sin\alpha\,(FU\cos\alpha - FW\sin\alpha) \quad \text{at} \quad s = L \tag{11.3}$$

in which

$$F = \frac{-P}{2\pi L \sin\alpha} \tag{11.4}$$

is the uniformly distributed force along the edge due to P.

A functional consisting of the preceding strain energy, potential energy, and constraint condition (1) may be constructed as

$$V = V_s + V_w - [\lambda\,(U\sin\alpha + W\cos\alpha)\,2\pi L\sin\alpha]_{s=L} \tag{11.5}$$

in which λ is the Lagrange's multiplier.

By using the strain-displacement relations, one obtains

$$e_s = U' + \frac{1}{2}(W')^2 \qquad e_\theta = \frac{U + W\cot\alpha}{s} \tag{11.6a,b}$$

$$K_s = -W'' \qquad K_\theta = -\frac{W'}{s} \tag{11.7a,b}$$

where a prime indicates the derivative with respect to s. Applying the variation method for a stationary value of V with respect to the displacements, U and W, the following two equations of equilibrium, respectively, are obtained

$$(sN_s)' - N_\theta = 0 \tag{11.8a}$$

$$(sN_sW')' - N_\theta\cot\alpha + (sM_s)'' - M_\theta' = 0 \tag{11.8b}$$

with the following boundary conditions for the present case

$$N_s = F\cos\alpha + \lambda\sin\alpha \quad \text{or} \quad U = 0 \quad \text{at} \quad s = L \tag{11.9a,b}$$
$$S_s = -F\sin\alpha + \lambda\cos\alpha \quad \text{or} \quad W = 0 \quad \text{at} \quad s = L \tag{11.10a,b}$$
$$W' = 0 \quad \text{or} \quad M_s = 0 \quad \text{at} \quad s = L \tag{11.11a,b}$$

and

$$U = 0 \quad \text{or} \quad N_s = 0 \quad \text{at} \quad s = L_1 \tag{11.12a,b}$$
$$W = 0 \quad \text{or} \quad S_s = 0 \quad \text{at} \quad s = L_1 \tag{11.13a,b}$$
$$W' = 0 \quad \text{or} \quad M_s = 0 \quad \text{at} \quad s = L_1. \tag{11.14a,b}$$

The shearing stress resultant, S_s, given in (11.10a) is defined as

$$S_s = N_s W' + (sW_s)'/s - M_\theta/s \tag{11.15}$$

This formulation reveals that conical shells like other shells have either the geometrical boundary conditions (11.9b, 11.10b, 11.11a, 11.12a, 11.13a and 11.14a), or natural boundary conditions (11.9a, 11.10a, 11.11b, 11.12b, 11.13b and 11.14b); (11.1) does not belong to either category. The Lagrange's multiplier, λ, physically is a constraint force at the deformable upper edge acting in the horizontal direction as shown in Figs. 11.1(b,c). This additional unknown, λ, is compensated by the additional constraint condition (11.1) which was treated for inclined bars for vibrations in Chap. 2, buckling in Chap. 8, and bending in Chap. 9. For the cases, $U = W = 0$ at $s = L$, and for cylindrical shells, $\alpha = 0$, the constraint condition vanishes automatically.

Using the generalized Hooke's law, one has

$$e_s = (N_s - \mu N_\theta)/Et \quad e_\theta = (N_\theta - \mu N_s)/Et \tag{11.16a,b}$$
$$M_s = D(K_s + \mu K_\theta) \quad M_\theta = D(K_\theta + \mu K_s) \tag{11.17a,b}$$

in which

$$D = Et^3/12(1-\mu^2) \tag{11.17c}$$

E = the modulus of elasticity; t = the shell thickness; and μ = Poisson's ratio.

When Airy's stress function, ψ, is introduced and defined as

$$N_s = \psi'/s \quad N_\theta = \psi'' \tag{11.18a,b}$$

then (11.8a) is satisfied. By using (11.6, 11.7, 11.16, 11.17 and 11.18), (11.8b) may be presented in the following form:

$$D\nabla^2\nabla^2 W = \left(W''\psi' + W'\psi'' - \psi''\cot\alpha\right)/s \tag{11.19}$$

in which

$$\nabla^2 = ()'' + ()'/s \tag{11.20}$$

By using (11.16 and 11.18), the compatibility equation [48]

$$2e_s'' + 2e_\theta' - e_s' = W'' \cot\alpha - W'W'' \tag{11.21}$$

becomes

$$\nabla^2\nabla^2\psi = \left[EtW''\left(\cot\alpha - W'\right)\right]/s \tag{11.22}$$

Equations (11.19 and 11.22) are the present version of the Donnell-type equations for conical shells [42].

11.2 Membrane and Bending Solutions

The solutions of conical shells may be separated into two parts: membrane and bending, which are indicated by superscripts, m and b, respectively. For instance

$$W = W^m + W^b \quad \psi = \psi^m + \psi^b \tag{11.23a,b}$$

Letting $D = 0$ and omitting the nonlinear terms in (11.19 and 11.22) result in the following two membrane equations for W^m and ψ^m, respectively.

$$(\psi^m)'' = 0 \qquad \nabla^2\nabla^2\psi^m = \frac{1}{s}\left[Et\cot\alpha\left(W^m\right)''\right] \tag{11.24a,b}$$

whose solutions, combined with (11.6, 11.16 and 11.18), are

$$\psi^m = A_1 s \tag{11.25}$$

$$W^m = \left[\tan\alpha\left(-A_1\ln s + A_2\right)\right]/Et \tag{11.26}$$

$$N_\theta^m = 0 \tag{11.27}$$

$$N_s^m = A_1/s \tag{11.28}$$

$$U^m = \left[A_1\left(\ln s - \mu\right) - A_2\right]/Et \tag{11.29}$$

in which A_1 and A_2 are constants of integration.

The bending effects for thin shells are confined in the near edge-zone. Therefore, the boundary-layer approach used for the plate in the last chapter is employed here.

A boundary-layer coordinate, η, is introduced at the small edge of the cone:

$$\eta = \left[(s/L) - 1\right]/k \tag{11.30}$$

in which

$$k = \left[(t/L)\tan\alpha/C_\mu\right]^{\frac{1}{2}} \quad C_\mu = \left[3\left(1-\mu^2\right)\right]^{\frac{1}{2}} \tag{11.31a,b}$$

Let

$$W^b(\eta) = t\left[W_0^b(\eta) + kW_1^0(\eta) + \ldots\right] \tag{11.32a}$$

$$\psi^b(\eta) = Et^3\left[\psi_0^b(\eta) + k\psi_1^b(\eta) + \ldots\right] \tag{11.32b}$$

Substituting (11.32) into (11.23) then into (11.19) and (11.22), with the variable s changed to η, and subtracting (11.24) from the resultant equations yield a set of equations associated with different orders of k.

The set with the lowest (zero) order of k has the form:

$$W^b_{0,\eta\eta\eta\eta} - 4TW^b_{0,\eta\eta} + 4C_\mu \psi^b_{0,\eta\eta} = 0 \tag{11.33a}$$

$$W^b_{0,\eta\eta} = C_\mu \psi^b_{0,\eta\eta\eta\eta} \tag{11.33b}$$

in which

$$T = C_\mu \tan\alpha N^m_s L/Et^2 \tag{11.34}$$

The N^m_s in the last equation is evaluated near the small edge, at where by (11.30) $\eta k \ll 1$. Thus $s \cong L$ and $N^m_s \cong A_1/L$ may be considered as constant. Therefore, (11.33) can be combined into one equation

$$W^b_{,\eta\eta\eta\eta} - 4TW^b_{,\eta\eta} + 4W^b = 0 \tag{11.35}$$

in which and thereafter, the subscript 0 is dropped for simplicity because only the lowest order solutions are sought.

Let

$$\beta = (1+T)^{1/2} \qquad \phi = (1-T)^{1/2} \tag{11.36a,b}$$

Then the solution of (11.35) for the boundary-layer is

$$W^b = e^{-\beta\eta}\left(B_1 \cos\phi\eta + B_2 \sin\phi\eta\right) \tag{11.37}$$

in which B_1 and B_2 are constants of integration. From (11.33b)

$$\psi^b = e^{-\beta\eta}\left[B_1\left(T\cos\phi\eta - \phi\beta\sin\phi\eta\right) + B_2\left(\phi\beta\cos\phi\eta + T\sin\phi\eta\right)\right] \tag{11.38}$$

Using (11.18, 11.7 and 11.17), one has

$$N^b_\theta = Et^2 e^{-\beta\eta}\left(B_1\cos\phi\eta + B_2\sin\phi\eta\right)\frac{\cot\alpha}{L} \tag{11.39}$$

$$M^b_s = -\frac{2Dt}{L^2k^2}e^{-\beta\eta}\left[\left(TB_1 - \phi\beta B_2\right)\cos\phi\eta + \left(\phi\beta B_1 + TB_2\right)\sin\phi\eta\right] \tag{11.40a}$$

$$M^b_\theta = \mu M^b_s \tag{11.40b}$$

Note that since N^b_s is of the order of k^1 as compared with N^b_θ of the order k^0, N^b_s has been discarded.

For the bottom end, there is another set of boundary-layer solutions, for which

$$\hat{\eta} = \left(1 - \frac{s}{L_1}\right)\frac{1}{\hat{k}} \qquad \hat{k} = k\left(\frac{L}{L_1}\right)^{1/2} \tag{11.41a,b}$$

The solutions may be obtained by adding a caret above the corresponding solutions in (11.37, 11.38, 11.39 and 11.40). For instance

$$\hat{W}^b(\hat{\eta}) = e^{-\hat{\beta}\hat{\eta}}\left(\hat{B}_1\cos\hat{\phi}\hat{\eta} + \hat{B}_2\sin\hat{\phi}\hat{\eta}\right) \tag{11.42}$$

in which $\hat{B}_1$ and $\hat{B}_2$ are another two constants of integration and

$$\hat{T} = T\frac{L}{L_1} \quad \hat{\beta} = \left(1 + \hat{T}\right)^{1/2} \quad \hat{\phi} = \left(1 - \hat{T}\right)^{1/2} \tag{11.43a-c}$$

Combining (11.6a) and (11.16a), one has

$$U' = \frac{N_s - \mu N_\theta}{Et} - \frac{1}{2}(W')^2 \tag{11.44}$$

Using the solutions of (11.37) and (11.39) and integrating the resulted equation yield a solution for $U^b(\eta)$. At the other end, there is $\hat{U}^b(\hat{\eta})$. The two constants of integration involved are determined by two conditions:

$$\hat{U}^b(0) = 0 \tag{11.45a}$$

$$U^b(\infty) = \hat{U}^b(\infty) \tag{11.45b}$$

The final form of U^b is given as

$$\begin{aligned}\frac{U^b(\eta)}{t} = \frac{C_\mu k \cot\alpha}{8} & ((4\mu/C_\mu)\left\{e^{-\beta\eta}\left[(\beta B_1 + \phi B_2)\cos\phi\eta\right.\right. \\ & + (-\phi B_1 + \beta B_2)\sin\phi\eta] \\ & \left. + (L/L_1)^{1/2}\left(\hat{\beta}\hat{B}_1 + \hat{\phi}\hat{B}_2\right)\right\} \\ & + (L/L_1)^{1/2}\left[\hat{\beta}\left(\hat{B}_1^2 - \hat{B}_2^2\right) - 2\hat{\phi}\hat{B}_1\hat{B}_2 + 2\left(\hat{B}_1^2 + \hat{B}_2^2\right)/\hat{\beta}\right] \\ & + e^{-2\beta\eta}\left\{\left[\beta\left(B_1^2 - B_2^2\right) - 2\phi B_1 B_2\right]\cos 2\phi\eta\right. \\ & \left. + \left[\phi\left(B_1^2 - B_2^2\right) + 2\beta B_1 B_2\right]\sin 2\phi\eta + 2\left(B_1^2 + B_2^2\right)/\beta\right\})\end{aligned} \tag{11.46}$$

The complete solutions are

$$W = W^m + W^b + \hat{W}^b \tag{11.47a}$$

$$U = U^m + U^b + \hat{U}^b \tag{11.47b}$$

$$N_s = N_s^m \tag{11.48a}$$

$$N_\theta = N_\theta^b + \hat{N}_\theta^b \tag{11.48b}$$

$$M_s = M_s^b + \hat{M}_s^b \tag{11.49a}$$

$$M_\theta = \mu M_s \tag{11.49b}$$

$$S_s = M_s' + N_s''' \left(W'^b + \hat{W}'^b\right) \tag{11.50}$$

11.3 Deformation

Consider the simply supported cone of Fig. 11.1(a). There are seven constants in the solution: six constants of integration and one Lagrange's multiplier to be determined by the following seven conditions:

$$N_s = \bar{N}_s \quad S_s = \bar{S}_s \quad M_s = 0 \quad \text{at} \quad s = L \tag{11.51a-c}$$
$$U = 0 \quad W = 0 \quad M_s = 0 \quad \text{at} \quad s = L_1 \tag{11.52a-c}$$

and

$$U \sin\alpha + W \cos\alpha = 0 \quad \text{at} \quad s = L \tag{11.53}$$

From (11.51a,b) and Fig. 11.1(b) or (11.9a and 11.10a), one has

$$\bar{N}_s = F \cos\alpha + \lambda \sin\alpha \tag{11.54a}$$
$$\bar{S}_s = -F \sin\alpha + \lambda \cos\alpha \tag{11.54b}$$

The evaluation of the constants may be simplified in the following way. Instead of F, let $\bar{N}_s$ defined in (11.54a) be considered as the independent loading parameter. By so doing, the Lagrange's multiplier, λ, and condition (11.51b) can also be avoided at this stage. Thus there are six conditions (11.51a,c; 11.52a-c; and 11.53) for the six constants. The results are:

$$A_1 = \frac{TEt^2 \cot\alpha}{C_\mu} \qquad A_2 = A_1 (\ln L - \mu) \tag{11.55a,b}$$

$$\hat{B}_1 = \frac{\mu T}{C_\mu} \qquad \hat{B}_2 = \frac{T\hat{B}_1}{\hat{\phi}\hat{\beta}} \tag{11.56a,b}$$

$$B_1 = \frac{-b + \left(b^2 - 4ac\right)^{1/2}}{2a} \qquad B_2 = \frac{TB_1}{\phi\beta} \tag{11.57a,b}$$

in which

$$a = \frac{t\,(2T+3)}{8Lk\beta^3} \tag{11.58a}$$

$$b = \cot\alpha + \frac{\mu t\,(1+2T)}{2kL\beta C_\mu} \tag{11.58b}$$

$$c = \frac{t\mu^2 T \left[4\left(1+2\hat{T}\right) + \frac{T\left(3+2\hat{T}\right)}{B_2}\right]}{8k\hat{\beta}\,(LL_1)^{1/2}\, C_\mu^2} - \frac{\mu T}{C_\mu \tan\alpha} \tag{11.58c}$$

and from (11.34 and 11.51a)

$$T = \frac{C_\mu L}{Et^2} \bar{N}_s \tan\alpha \tag{11.59}$$

The minus sign preceding the radical of (11.57a) was dropped because of the following consideration: When $\mu = 0, c, B_1$ to $\hat{B}_2$ from (11.56, 11.57 and 11.58c) all vanish. So there is no bending solution.

By use of these solutions, (11.50) yields the shearing stress resultant, S_s, which evaluated at $\eta = 0$ has the form

$$\bar{S}_s = \frac{-B_1 Et^3\,(t+0.5)}{k\beta L^2 C_\mu \tan\alpha} \tag{11.60}$$

Solving (11.54) for F and λ

$$F = \bar{N}_s \cos\alpha - \bar{S}_s \sin\alpha \tag{11.61}$$

$$\lambda = \bar{N}_s \sin\alpha + \bar{S}_s \cos\alpha \tag{11.62}$$

which are the equilibrium conditions at the smaller end between external and internal forces as shown in Fig. 11.1(b).

When the load, P, is normalized with respect to the buckling load of cylindrical shells, P_{cy} [48], and let

$$\tau = \frac{P}{P_{cy}} = \frac{PC_\mu}{2\pi Et^2} \tag{11.63}$$

Equation (11.4) yields

$$F = -\frac{\tau Et^2}{LC_\mu \sin\alpha} \tag{11.64}$$

Using (11.59, 11.60, 11.61 and 11.64), one has

$$\tau = -\left[T + \frac{B_1 t\,(T+0.5)\sin\alpha}{Lk\beta}\right]\cos^2\alpha \tag{11.65}$$

The lower edge of the cone is considered being fixed in space, and the upper edge will experience a head-shortening, δ, as end B moves to B' as shown in Fig. 11.1(c), thus

$$\delta = \frac{U\cos\alpha - W\sin\alpha}{t} \quad \text{at} \quad s = L \tag{11.66}$$

By using the obtained solutions for U and W

$$\delta = \frac{T}{C_\mu}\left[\ln\left(\frac{L}{L_1}\right)\csc\alpha + \mu\sin\alpha\right] - B_1\sin\alpha + \frac{U^b(0)\cos\alpha}{t} \tag{11.67}$$

The two (11.65 and 11.67) may be used to plot a force versus head-shortening curve by considering $\bar{N}_s$ or T of (11.59) as a parameter. Such a typical curve is given in Fig. 11.2 marked by τ_0. The maximum value denoted by τ_{0c} is the buckling load of the cone. Let the buckling load be normalized by $\cos^2\alpha$ and denoted as

$$Q_0 = \frac{\tau_{0c}}{\cos^2\alpha} \tag{11.68}$$

Figure 11.3, in which $w = W/t$ and $u = U/t$, presents two elastic curves of meridional elements of cones of $\alpha = 15°$ and $75°$ at the respective τ_{0c}. The elastic curve of $\alpha = 15°$ reveals that the angular rotation at the loaded edge is considerable. A study on the effect of this angle change to the buckling strength of the cones is made in the following section.

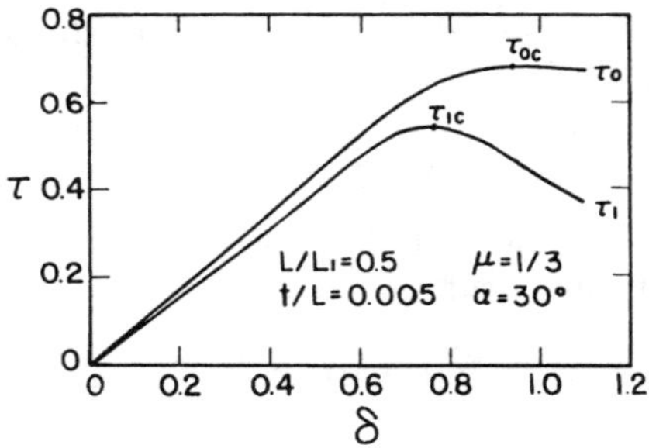

Fig. 11.2. Force vs. head-shortening curves

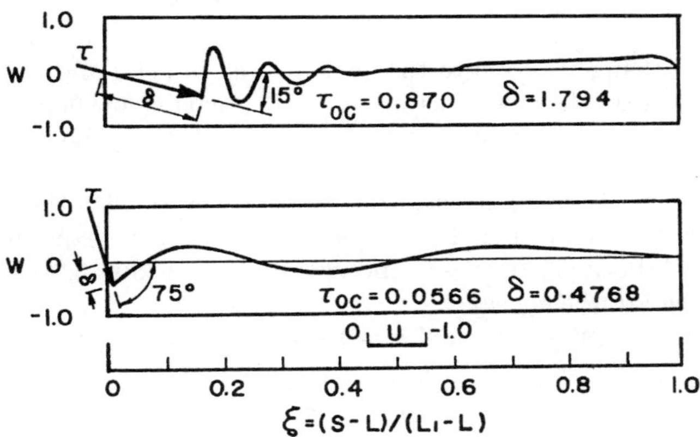

Fig. 11.3. Elastic curves at buckling loads evaluated at undeformed states

11.4 Effect of Angle Change

The buckling strength of structures is supposedly evaluated at the deformed states. As it was stated earlier that if an axial or in-plane load is applied on columns, plates, and cylindrical shells, the geometrical configurations of these structures, theoretically, do not change during the loading process until buckling takes place. But for conical shells, because of condition (11.1), a lateral displacement, W, will be induced by in-plane displacement, U, even in the pre-buckling state. In turn, the angle, α, at the smaller end will be changed to α_1 as shown in Fig. 11.1(c),

$$\alpha_1 = \alpha + W' \quad \text{at} \quad s = L \tag{11.69}$$

which should be used in such equations as (11.61 and 11.62). Since N_s and S_s are derived at the undeformed state and at deformed states, they should be replaced by, respectively,

$$N_{s1} = N_s / (1 + e_s) \quad S_{s1} = S_s \tag{11.70a,b}$$

There is no change in the latter because, for thin shells, the shearing strain normal to the middle surface has been neglected. At the smaller edge

$$N_{s1} = \bar{N}_{s1} \qquad S_{s1} = \bar{S}_{s1} \quad \text{at} \quad s = L \tag{11.71a,b}$$

Then the equilibrium equation, (11.61), is evaluated at the deformed state as shown in Fig. 11.1(c), this yields

$$F = \bar{N}_{s1} \cos\alpha_1 - \bar{S}_{s1} \sin\alpha_1 \tag{11.72}$$

Accordingly, τ changes to

$$\tau_1 = -\left[T(1+e_s) + \frac{B_1 t\,(T+0.5)\tan\alpha_1}{Lk\beta}\right] \cos\alpha_1 \cot\alpha_1 \sin\alpha \tag{11.73}$$

Note that no change is given to the angle α used in (11.4 and 11.66). This is why the angle α in the sine function of (11.73) remains unchanged.

Using this τ_1 value, another force versus head-shortening curve may be constructed, such as the τ_1 curve shown in Fig. 11.2. The buckling load, τ_{1c}, is lower than τ_{0c}. Defining a set of Q_1 versus α curves with the t/L ratio as a parameter is shown in Fig. 11.4. Presented are also the experiment results obtained by Arbocz [5]. In these experiments, the buckling also started from the smaller end of the cones.

Elastic curves for the cones of $\alpha = 15°$ and $75°$ at $\tau = \tau_{1c}$ are depicted in Fig. 11.5.

$$Q_1 = \frac{\tau_{1c}}{\cos^2\alpha} \tag{11.74}$$

11.5 Discussion of the Results

From the numerical results obtained, the following observations are made:

1. The boundary-layer technique is effective for thin shells. The thinner is the shell, the better results will be. This fact is confirmed by the Arbocz' experiments of Group I which was the thinnest group ever tested, and the results are most close to the analytic predictions in Fig. 11.4.
2. The elastic curves of Figs. 11.3 and 11.5 indicate that the effect of angle change from α to α_1 for cones with small α's is larger than for cones of large α's as shown in these figures.
3. Figures 11.3 and 11.5 also indicate that for a cone with small α (perhaps $<$ $30°$), buckling is confined at the smaller end; such buckling may be known as local buckling and was observed in experiments [5,25]. However, for a cone with large α , the buckling effect extends to the whole cone.
4. The localized buckling apparently weakens the buckling strength of these cones. For a cone with smaller $\alpha (> 0)$, the buckling may be localized, thus the cone will get weaker. However, there are two different limiting cases of $\alpha = 0$.

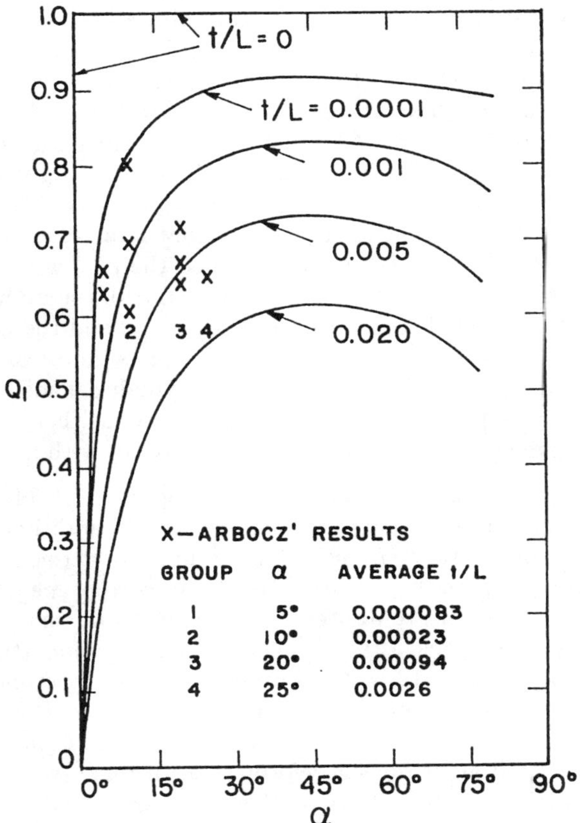

Fig. 11.4. Normalized buckling loads

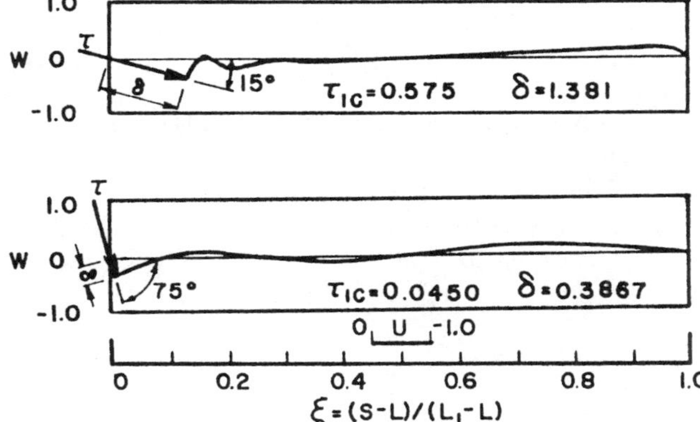

Fig. 11.5. Elastic curves at buckling loads evaluated at deformed states

5. First case, consider the cone of Fig. 11.1(a) with a fixed distance $OB = L$, (hereafter thickness t is considered as a constant.) Thus the t/L ratio is a constant. Now let α decrease. Physically, this is the case of cones generated by rotating element AB about apex 0. As $\alpha \to 0$, Fig. 11.4 shows $Q_1 \to 0$, which is consistent with the fact that at the limiting case of $\alpha = 0$, the cone becomes a geometrical line which has little buckling strength.
6. Second case, consider the cone of Fig. 11.1(a) with a fixed R_1. Let element AB rotate about end B. In this case, the distant L increases as α decreases. When $t/L \to 0$ and $L \to \infty$, $\alpha \to 0$. When the t/L ratio is decreasing, the Q_1 curves in Fig. 11.4 shift upward. As $t/L = 0$, the limiting Q_1 curve forms a right angle made of the vertical line $\alpha = 0$ and the horizontal line $Q_1 = 1$ as shown in Fig. 11.4. When α approaches zero from the positive side, $\alpha = 0$ and $Q_1 = 1$, the cone becomes a cylindrical shell.
7. It is noted that the present asymptotic solutions are neither valid for conical shells with large t/L ratio when the shell is thick nor with α closing to $\pi/2$.

The present results indicate that the buckling strength of conical shells mainly depends on the condition of the smaller end. In addition to the vertex angle, α, the distance ratio, t/L, also plays an important role.

The up to 15% differences between the Arbocz's experimental results and Q_1 curves in Fig. 11.4, for group other than the first one, may be attributed to: (1) initial imperfections, (2) asymmetrical deformation (the number of circumferential waves were 12 to 17 as observed in the experiments [5]), and (3) the use of a strain-softening material (copper) in the experiments.

When the smaller end of a cone is fixed against rotation, the angle, α, will retain its original angle during deformation. In this case, however, there is a concentration of bending stress at that end. An inelastic hinge may be developed. If this happens, the angle α will also change. Inelastic behavior, however, is beyond the scope of the present investigation.

Exercises

1. Using the variation method, derive (11.8 to 11.14).
2. Derive (11.19 and 11.22).
3. Show that (11.37 and 11.38) satisfy (11.33).
4. Determine τ_1 and τ_0 versus δ curves for a cone with the same constants shown in Fig. 11.2, except $\alpha = 30°$.

12

Conical Shells of Linearly Varying Thickness Subjected to Normal Loads

The equilibrium equations for a surface element of a conical shell in the framework of generalized plane stress of linear theory of elasticity are given in [19]. These equations may be applied to linearly varying thickness as well as to constant thickness of conical shells.

For linearly varying thickness cones, the elastic relations between stress resultants and displacements are also given along with general approach to solve the basic equations in [19]. A set of homogenous equations in terms of three displacement components are solved by the method of separation of variables; in turn, the solutions are depending on an eighth degree characteristic equation. In the present study, the characteristic equation is rearranged so that it can be solved in an explicit form.

A set of asymptotic complete solutions is presented. The solutions have two parts: membrane and bending. Both parts are coupled by normal displacements. Two numerical examples of conical shells subjected to normal loads are given.

12.1 Basic and Characteristic Equations

The equilibrium equations for a middle surface element of a conical shell from [19] are

$$\begin{aligned}
(sN_s)^{\cdot} + N'_{\theta s}\sec\alpha - N_\theta &= -P_s s \\
(sN_{s\theta})^{\cdot} + N'_\theta \sec\alpha + N_{\theta s} - Q_\theta \tan\alpha &= -P_\theta s \\
N_\theta \tan\alpha + Q'_\theta \sec\alpha + (sQ_s)^{\cdot} &= P_r s \\
(sM_s)^{\cdot} + M'_{\theta s}\sec\alpha - M_\theta &= sQ_s \\
(sM_{s\theta})^{\cdot} + M'_\theta \sec\alpha + M_{\theta s} &= sQ_\theta \\
s\,(N_{\theta s} - N_{s\theta}) &= M_{s\theta}\tan\alpha
\end{aligned} \tag{12.1a-f}$$

in which the θ and s are the circumferential and meridional coordinates of the middle surface; the N_s, N_θ, $N_{s\theta}$, $N_{\theta s}$, M_s, M_θ, $M_{s\theta}$, $M_{\theta s}$, and Q_s, Q_θ are stress, moment, and shear stress resultants per unit length; P_s, P_θ, and P_r are surface loads per unit area in the directions indicated by the subscripts, respectively; the r is the coordinate in the normal to the surface direction, outward positive; a dot represents differentiation with respect to s and a prime with respect to θ; and α is the angle between the shell surface and vertical central line of the cone.

Equation (12.1f) does not involve differentiation, hence it may be considered later. When the lateral normal load P_r is considered only, P_s and $P_\theta = 0$. Substituting Q_s and Q_θ from (12.1d,e) into (12.1b,c), one has three equations. With the order of (12.1a,b) exchanged, one has

$$\begin{aligned} s\left(sN_{s\theta}\right)^{\cdot} + sN_\theta' \sec\alpha + sN_{\theta s} - \left(sM_{s\theta}\right)^{\cdot} \tan\alpha & \\ -M_{\theta s}\tan\alpha - M_\theta' \tan\alpha\sec\alpha &= 0 \\ \left(sN_s\right)^{\cdot} + N_{\theta s}' \sec\alpha - N_\theta &= 0 \qquad (12.2\text{a-c}) \\ sN_\theta \tan\alpha + s\left(sM_s\right)^{\cdot\cdot} + \left(sM_{s\theta}'\right)^{\cdot}\sec\alpha + \left(sM_{\theta s}'\right)^{\cdot}\sec\alpha & \\ + M_\theta'' \sec^2\alpha - sM_\theta^{\cdot} &= P_r s^2 \end{aligned}$$

For a conical shell with linearly varying thickness, the thickness h is proportion to s, so that

$$h = \delta s \tag{12.3}$$

where δ is a proportional small constant for thin shells. The elastic law yields the following relationships between stress resultants and displace components

$$\begin{aligned} N_s &= D\left[v^{\cdot} + \frac{\nu}{s}\left(\frac{u'}{\cos\alpha} + v + w\tan\alpha\right)\right] - K\frac{w^{\cdot\cdot}}{s}\tan\alpha \\ N_\theta &= D\left[\frac{1}{s}\left(\frac{u'}{\cos\alpha} + v + w\tan\alpha\right) + \nu v^{\cdot}\right] \\ &\quad + K\left[\frac{v}{s^3}\tan\alpha + \frac{w}{s^3}\tan^2\alpha + \frac{w''}{s^3\cos^2\alpha} + \frac{w^{\cdot}}{s^2}\right]\tan\alpha \\ N_{s\theta} &= D\frac{1-\nu}{2}\left[u^{\cdot} - \frac{u}{s} + \frac{v'}{s\cos\alpha}\right] \\ &\quad + K\frac{1-\nu}{2}\left[\frac{u^{\cdot}}{s^2} - \frac{u}{s^3} - \frac{w'^{\cdot}}{s^2\sin\alpha} + \frac{w'}{s^3\sin\alpha}\right]\tan^2\alpha \\ N_{\theta s} &= D\frac{1-\nu}{2}\left[u^{\cdot} - \frac{u}{s} + \frac{v'}{s\cos\alpha}\right] \qquad (12.4\text{a-h}) \\ &\quad + K\frac{1-\nu}{2}\left[\frac{v'}{s^3\cos\alpha} + \frac{w'^{\cdot}}{s^2\sin\alpha} - \frac{w'}{s^3\sin\alpha}\right]\tan^2\alpha \\ M_s &= K\left[w^{\cdot\cdot} - \frac{v^{\cdot}}{s}\tan\alpha + \nu\left(\frac{w''}{s^2\cos^2\alpha} + \frac{w^{\cdot}}{s} - \frac{u'\tan\alpha}{s^2\cos\alpha}\right)\right] \\ M_\theta &= K\left[\frac{w''}{s^2\cos^2\alpha} + \frac{w^{\cdot}}{s} + \frac{w}{s^2}\tan\alpha + \frac{v}{s^2}\tan\alpha + \nu w^{\cdot\cdot}\right] \end{aligned}$$

$$M_{s\theta} = K(1-\nu)\left[\frac{w'^{\cdot}}{s} - \frac{w'}{s^2} - \frac{u^{\cdot}}{s}\tan\alpha + \frac{u}{s^2}\sin\alpha\right]\frac{1}{\cos\alpha}$$

$$M_{\theta s} = K(1-\nu)\left[\frac{w'^{\cdot}}{s} - \frac{w'}{s^2} - \frac{u^{\cdot}}{2s}\sin\alpha + \frac{u}{2s^2}\sin\alpha + \frac{v'}{2s^2}\tan\alpha\right]\frac{1}{\cos\alpha}$$

in which u, v, and w are circumferential, meridional, and normal displacement components, respectively; ν is Poison's ratio; and

$$D = Eh/(1-v^2) \qquad K = Eh^3/12(1-v^2) \tag{12.5a,b}$$

where E is modulus of elasticity. Letting

$$k = \delta^2/12 \tag{12.6}$$

and substituting the elastic law (12.4) into (12.2) result in three equilibrium equations in terms of three displacements:

$$\begin{aligned}
&\frac{1-\nu}{2}s^2u^{\cdot\cdot} + u''\sec^2\alpha + (1-\nu)\,su^{\cdot} - (1-\nu)\,u + \frac{1+\nu}{2}sv'^{\cdot}\sec\alpha \\
&\quad + (2-\nu)\,v'\sec\alpha + w'\tan\alpha\sec\alpha + k\Big[3\,(1-\nu)\left(s^2u^{\cdot\cdot}\tan\alpha/2 + su^{\cdot}\tan\alpha\right. \\
&\quad \left. -u\tan\alpha - sw'^{\cdot}\sec\alpha + w'\sec\alpha\right) - \frac{3-\nu}{2}s^2w'^{\cdot\cdot}\sec\alpha\Big]\tan\alpha = 0 \\[2ex]
&\frac{1+\nu}{2}su'^{\cdot}\sec\alpha - \frac{3}{2}(1-\nu)\,u'\sec\alpha + s^2v^{\cdot\cdot} + \frac{1-\nu}{2}v''\sec^2\alpha \\
&\quad + 2sv^{\cdot} - (1-\nu)\,(v + w\tan\alpha) + \nu sw^{\cdot}\tan\alpha \\
&\quad + k\left[\frac{1-\nu}{2}\left(v''\tan\alpha\sec^2\alpha + sw''^{\cdot}\sec\alpha\right) - v\tan\alpha - s^3w^{\cdot\cdot\cdot}\right. \\
&\quad \left. - 3s^2w^{\cdot\cdot} - sw^{\cdot} - w\tan^2\alpha - \frac{3-\nu}{2}w''\sec^2\alpha\right]\tan\alpha = 0
\end{aligned} \tag{12.7a-c}$$

$$\begin{aligned}
&[u'\sec\alpha + \nu sv^{\cdot} + v + w\tan\alpha]\tan\alpha + k\left[-\frac{3-\nu}{2}s^2u'^{\cdot\cdot}\sec\alpha\right. \\
&\quad - (3+\nu)\,su'^{\cdot}\sec\alpha + (3-5\nu)\,u'\sec\alpha - s^3v^{\cdot\cdot\cdot} - 6s^2v^{\cdot\cdot} \\
&\quad \left. + (2-\nu)\,v''\sec^2\alpha - 7sv^{\cdot} - \nu\left(1-\tan^2\alpha\right) + \frac{1-\nu}{2}sv''^{\cdot}\sec^2\alpha\right]\tan\alpha \\
&\quad + k\left[s^4w^{\cdot\cdot\cdot\cdot} + 2s^2w''^{\cdot\cdot}\sec^2\alpha + w''''\sec^4\alpha + 8s^3w^{\cdot\cdot\cdot} + 4sw''^{\cdot}\sec^{2\alpha}\right. \\
&\quad + (11+3\nu)\,s^2w^{\cdot\cdot} + 2w''\tan^2\alpha\sec^2\alpha - (5-6\nu)\,w''\sec^2\alpha \\
&\quad \left. -2\,(1-3\nu)\,sw^{\cdot} - w\left(1-\tan^2\alpha\right)\tan^2\alpha\right] = sP_r\left(1-\nu^2\right)/E\delta
\end{aligned}$$

Introducing a non-dimensional variable

$$y = \sqrt{\frac{s}{L}} \tag{12.8}$$

(Note: In this and next two chapters $L > L_1$.) The displacement functions may now be assumed in the forms:

$$u = A_n y^{\lambda_n - 1} \frac{\sin n\pi\theta}{\cos \theta_1}$$
$$v = B_n y^{\lambda_n - 1} \frac{\cos n\pi\theta}{\sin \theta_1} \quad (12.9\text{a-c})$$
$$w = C_n y^{\lambda_n - 1} \frac{\cos n\pi\theta}{\sin \theta_1}$$

in which A_n, B_n, C_n, and λ_n are constants to be determined, $n = 1, 2, 3, \ldots$. Physically speaking, the upper set of the sinusoidal functions in (12.9) is for a complete cone, and the lower set is for a segment of cone which is bounded by $\theta = 0$ and θ_1 ($<2\pi$) and $s = L_1$ and L. The segment has two simply supported generator edges which have the following boundary conditions:

$$w = 0 \quad v = 0 \quad N_\theta = 0 \quad \text{and} \quad M_\theta = 0 \quad \text{at } \theta = 0 \text{ and } \theta_1 \qquad (12.10\text{a-d})$$

The reactions along the two generator edges are given by

$$S_\theta = Q_\theta + \frac{\partial M_{s\theta}}{\partial S} \quad \text{at } \theta = 0 \text{ and } \theta_1 \qquad (12.10\text{e})$$

where S_θ is transverse shearing force at a section perpendicular to the θ direction. The shearing force Q_θ may be obtained from (12.1e).

Letting

$$P_r = p_{rn}(y) \frac{\cos n\pi\theta}{\sin \theta_1} \qquad (12.11)$$

and substituting this load function and assumed displacements (12.9) into (12.7) yield the following three equations:

$$d_{11}A_n + d_{12}B_n + d_{13}C_n = 0$$
$$d_{21}A_n + d_{22}B_n + d_{23}C_n = 0 \qquad (12.12\text{a-c})$$
$$d_{31}A_n + d_{32}B_n + d_{33}C_n = Lp_{rn}(y)\, y^{3-\lambda_n} \frac{1-\nu^2}{E\delta}$$

When the upper set of sinusoidal functions in (12.9) is used, the coefficients d_{mn} $(m, n = 1, 2, 3)$ are

$$d_{11} = \frac{1-\nu}{8}\left(1 + 3k\tan^2\alpha\right)\left(9 - \lambda^2\right) + \omega^2 \sec^2\alpha$$
$$d_{12} = \frac{1}{4}\left[7 - 5\nu + (1+\nu)\lambda\right]\omega\sec\alpha$$
$$d_{13} = \left\{1 + \frac{k}{8}\left[3(9 - 11\nu) + 8\nu\lambda - (3-\nu)\lambda^2\right]\right\}\omega\tan\alpha\sec\alpha$$
$$d_{22} = \frac{1}{4}\left(1 - \lambda^2\right) + (1-\nu)\left(1 + \frac{1}{2}\omega^2\sec^2\alpha\right)$$

$$+k\tan^2\alpha\left(1+\frac{1-\nu}{2}\omega^2\sec^2\alpha\right) \tag{12.13a-f}$$

$$\begin{aligned} d_{23} &= \frac{1}{2}\tan\alpha\,[2-\nu-\nu\lambda] \\ &-\frac{1}{k}\tan\alpha\,\big[1-8\tan^2\alpha+2\,(7-3\nu)\,\omega^2\sec^2\alpha \\ &-\left(3+2\,(1-\nu)\,\omega^2\sec^2\alpha\right)\lambda+3\lambda^2-\lambda^3\big] \\ d_{33} &= \tan^2\alpha+\frac{1}{16}k\,\big[(13-12\nu)-16\left(1-\tan^2\alpha\right)\tan^2\alpha \\ &+8\left(11-12\nu-4\tan^2\alpha\right)\omega^2\sec^2\alpha+16\omega^4\sec^4\alpha \\ &-2\left(7-6\nu+4\omega^2\sec^2\alpha\right)\lambda^2+\lambda^4\big] \end{aligned}$$

where

$$\omega = n\pi/\theta_1 \tag{12.13g}$$

If the lower set of sinusoidal functions is used, a minus $(-)$ sign shall be added to d_{12} and d_{13}. The expressions for d_{21}, d_{31}, and d_{32} are obtained by replacing λ with $-\lambda$ in d_{12}, d_{13}, and d_{23}, respectively.

The complete solution of (12.12) consists of two parts: homogenous and particular. The homogenous solution is considered first.

In order to have a non-trivial solution of A_n, B_n and C_n, the determinant of the coefficients must vanish. This results in an eighth degree characteristic equation for λ_n. Neglecting the terms of second and higher powers of k as it has been done in the derivation of elastic law (12.4), the characteristic equation is obtained in the following form:

$$G\left[\lambda_n^4-10\lambda_n^2+9\right]+k\left[\lambda_n^8-g_6\lambda_n^6+g_4\lambda_n^4-g_2\lambda_n^2+g_0\right]=0 \tag{12.14a}$$

in which

$$G=16(1-\nu^2)\tan^2\alpha \tag{12.14b}$$

and

$$\begin{aligned} g_6 &= 4\,(7-4\nu)-8\nu\tan^2\alpha+16\omega^2\sec^2\alpha \\ g_4 &= 2\,\big[\left(127-136\nu+24\nu^2\right) \\ &-4\,(8-3\nu)\tan^2\alpha+8\left(4-3\nu^2\right)\tan^4\alpha\big] \\ &+16\left[17-12\nu-6\tan^2\alpha\right]\omega^2\sec^2\alpha+96\omega^4\sec^4\alpha \\ g_2 &= 4\,\big[\left(203-316\nu+120\nu^2\right) \\ &-2\,(80-61\nu)\tan^2\alpha+40\left(4-3\nu^2\right)\tan^4\alpha\big] \\ &+16\,\big[(71-72\nu)-4\,(13-10\nu)\tan^2\alpha \\ &+8\,(2-\nu)\tan^4\alpha\big]\,\omega^2\sec^2\alpha \\ &+64\left[13-12\nu-2\,(4-\nu)\tan^2\alpha\right]\omega^4\sec^4\alpha+256\omega^6\sec^6\alpha \\ g_0 &= 9\left[(13-12\nu)\,(5-4\nu)-8\,(8-7\nu)\tan^2\alpha+16\left(4-3\nu^2\right)\tan^4\alpha\right] \end{aligned} \tag{12.14c-f}$$

$$
\begin{aligned}
&+ 16\left[\left(215 - 412\nu + 192\nu^2\right) + 2\left(89 - 172\nu + 96\nu^2\right)\tan^2\alpha\right. \\
&\left.+ 40\left(2-\nu\right)\tan^4\alpha\right]\omega^2\sec^2\alpha \\
&- 32\left[\left(81 - 184\nu + 96\nu^2\right) + 4\left(16 - 13\nu\right)\tan^2\alpha - 8\tan^4\alpha\right]\omega^4\sec^4\alpha \\
&+ 256\left[3 - 4\nu - 2\tan^2\alpha\right]\omega^6\sec^6\alpha + 256\omega^8\sec^8\alpha
\end{aligned}
$$

The coefficients g_6 to g_0 given in (12.14c-f) are the same as those given in [19] for the terms with the parameter k^1. The terms with k° are placed in the first bracket in (12.14a), so that the characteristic equation is consistent with the elastic law of (12.4).

In view of the approximation made in the derivation of (12.14a), the following approximate method for the solution of (12.14a) is introduced. Letting

$$\lambda_n^2 = X_{0n} + kX_{1n} \tag{12.15a}$$

and substituting (12.15a) into (12.14a) result in a sequence of equations associated with various powers of k. The equations associated with the two lowest powers of k, k^0 and k^1, are, respectively,

$$X_{0n}^2 - 10X_{0n} + 9 = 0 \tag{12.15b}$$

and

$$X_{0n}^4 - g_6X_{0n}^3 + g_4X_{0n}^2 - g_2X_{0n} + g_0 + 2G\left(X_{0n} - 5\right)X_{1n} = 0 \tag{12.15c}$$

which give two sets of X_{0n} and X_{1n}. Then (12.15a) provides two roots of λ_n^2, which, in turn, give four roots of λ_n

$$\lambda_{\substack{1n\\2}} = \pm\left(1 + k\frac{1 - g_6 + g_4 - g_2 + g_0}{8G}\right)^{1/2} \tag{12.16a}$$

$$\lambda_{\substack{3n\\4}} = \pm\left(9 - k\frac{9^4 - 9^3g_6 + 9^2g_4 - 9g_2 + g_0}{8G}\right)^{1/2} \tag{12.16b}$$

Substituting the determined roots of λ_n^2 into (12.14a) yields a quadratic equation of λ_n^2 which gives the other four roots of λ_n:

$$\lambda_{\substack{56n\\78}} = \pm\left\{\frac{1}{2}\left(g_6 - P - Q\right) \pm i\left[\frac{1}{PQ}\left(g_0 + \frac{9G}{k}\right) - \frac{1}{4}\left(g_6 - P - Q\right)^2\right]^{1/2}\right\}^{1/2} \tag{12.17a}$$

where

$$P = \lambda^2_{\substack{1n\\2}} \quad \text{and} \quad Q = \lambda^2_{\substack{3n\\4}} \tag{12.17b}$$

Hence the eight roots of λ_n are divided into two groups, four each. The first group is of real numbers; the other is of complex numbers.

The next step is to solve (12.12) for A_n and B_n in terms of C_n for each root of λ_n from any two of the homogeneous equations of (12.12). Eight constants C_n are determined by eight conditions at $y = \sqrt{L_1/L}$ and 1. For a segment of a cone, the boundary conditions along the generator edges are satisfied by the choice of sinusoidal functions of the angle θ. At the two circular edges, one has the following four boundary conditions at each edge.

For a built-in edge:

$$u = 0 \quad v = 0 \quad w = 0 \quad \text{and} \quad \partial w/\partial s = 0 \tag{12.18a-d}$$

and for a free edge:

$$N_s = 0 \quad M_s = 0 \quad S_s = 0 \quad \text{and} \quad T_s = 0 \tag{12.19a-d}$$

where

$$S_s = Q_s + \frac{1}{s}\frac{\partial M_{s\theta}}{\partial \vartheta} \sec\alpha \tag{12.20a}$$

$$T_s = N_{s\theta} - \frac{M_{s\theta}}{s} \tan\alpha \tag{12.20b}$$

are the transverse and tangential shearing forces at sections perpendicular to the s-direction, respectively. The shearing force Q_s can be obtained from (12.1d).

For a simply supported edge:

$$w = 0 \qquad M_s = 0 \tag{12.21a,b}$$

$$N_s = 0 \quad \text{or} \quad v = 0 \qquad T_s = 0 \quad \text{or} \quad u = 0 \tag{12.21c-f}$$

12.2 Asymptotic Solutions

As the parameter k approaches zero, the two groups of roots λ_n reach at the following asymptotic values:

$$\lambda_{\substack{1\\2}} = \pm 1 \qquad \lambda_{\substack{3\\4}} = \pm 3 \tag{12.22a,b}$$

$$\lambda_{\substack{5\\6}} = \rho\,(1 \pm i) \qquad \lambda_{\substack{7\\8}} = -\rho\,(1 \pm i) \tag{12.22c,d}$$

where

$$\rho \equiv \left| \left(\frac{1}{2}\right)^{1/2} \left(\frac{G}{k}\right)^{1/4} \right| \tag{12.23}$$

The subscript n has been and henceforth will be dropped for simplicity.

When the first group of λ, λ_j (j = 1, 2, 3 and 4), is substituted into the first two of (12.12) to eliminate A_i and B_i, and keeping only the leading terms of k, solutions of (12.7) assume the following forms:

$$\begin{aligned} u^1 &= \mp m \tan\alpha \left\{ \frac{C_1}{m^2-1} + \frac{C_2}{m^2-2(1-\nu)}\frac{1}{y^2} + \frac{C_3}{m^2}y^2 \right. \\ &\quad \left. + \frac{4(1+\nu)-m^2}{m^2(7-2\nu-m^2)}\frac{C_4}{y^4} \right\} \frac{\sin}{\cos}\frac{n\pi\theta}{\theta_1} \\ v^1 &= \tan\alpha \left\{ \frac{C_1}{m^2-1} + \frac{2C_2}{m^2-2(1-\nu)}\frac{1}{y^2} \right. \\ &\quad \left. + \frac{3C_4}{m^2-7+2\nu}\frac{1}{y^4} \right\} \frac{\cos}{\sin}\frac{n\pi\theta}{\theta_1} \\ w^1 &= \left\{ C_1 + C_2\frac{1}{y^2} + C_3 y^2 + C_4\frac{1}{y^4} \right\} \frac{\cos}{\sin}\frac{n\pi\theta}{\theta_1} \end{aligned} \tag{12.24a-c}$$

in which

$$m = \frac{n\pi}{\theta_1}\sec\alpha \tag{12.24d}$$

When the second group of λ, λ_i ($i = 5, 6, 7$, and 8), is used following the similar procedure and using some identities to convert the complex expressions into real functions, one obtains:

$$\begin{aligned} u^{11} &= \mp 2(2+\nu)\, m \tan\alpha \frac{1}{\rho^2}\frac{1}{y} \{ y^\rho [-C_5 \sin(\rho\ln y) + C_6\cos(\rho \ln y)] \\ &\quad + y^{-\rho}[C_7\sin(\rho\ln y) - C_8\cos(\rho\ln y)]\} \frac{\sin}{\cos}\frac{n\pi\theta}{\theta_1} \\ v^{11} &= -\nu\tan\alpha\frac{1}{\rho}\frac{1}{y}\{ y^\rho[(C_5-C_6)\cos(\rho\ln y) + (C_5+C_6)\sin(\rho\ln y)] \\ &\quad - y^{-\rho}[(C_7+C_8)\cos(\rho\ln y) - (C_7-C_8)\sin(\rho\ln y)]\} \frac{\cos}{\sin}\frac{n\pi\theta}{\theta_1} \\ w^{11} &= y^{-1}\{ y^\rho[C_5\cos(\rho\ln y) + C_6\sin(\rho\ln y)] \\ &\quad + y^{-\rho}[C_7\cos(\rho\ln y) + C_8\sin(\rho\ln y)]\} \frac{\cos}{\sin}\frac{n\pi\theta}{\theta_1} \end{aligned} \tag{12.25a-c}$$

It is noted that the solutions of the first group are simply those of membrane theory. Based on solutions (12.24 and 12.25), one may establish the orders of magnitude of the displacement components:

$$u^1, v^1, w^1, w^{11} = O\left(\frac{1}{\rho^0}\right) \tag{12.26a}$$

$$v^{11} = O\left(\frac{1}{\rho}\right) \quad \text{and} \quad u^{11} = O\left(\frac{1}{\rho^2}\right) \tag{12.26b,c}$$

It is assumed that the parameter m is limited to small values so that the differentiation with respect to the θ does not affect the order of magnitude. Thus n is limited to small numbers.

Due to u^1, v^1 and w^1, the magnitudes of the corresponding stresses N_s^1, N_θ^1 and $N_{\theta s}^1$ obtained by using (12.4) are also of the order of $1/\rho^0$ and the moments are of $1/\rho^3$ and higher. The orders of magnitude of the stresses due to u^{11}, v^{11}, and w^{11} are not quite obvious and will be examined as follows.

Changing the variable s to y according to (12.8) and then to η, such that

$$y = \eta^{1/\rho} \tag{12.27}$$

and from (12.25), one has

$$\begin{aligned} u = u^{11} &= \frac{1}{\rho^2} U \\ v = v^{11} &= \frac{1}{\rho} V \\ w = w^{11} &= \frac{1}{\rho^0} W \end{aligned} \tag{12.28a-c}$$

When the terms of the lowest order of $1/\rho$ only are retained, the elastic law (12.4) leads to the following expressions:

$$\begin{aligned} N_s^{11} &= \frac{E\delta}{1-\nu^2}\left[\frac{1}{2}\eta\frac{\partial V}{\partial \eta} + \nu W \tan\alpha\right] \\ N_\vartheta^{11} &= \frac{E\delta}{1-\nu^2}\left[W\tan\alpha + \nu\frac{1}{2}\eta\frac{\partial V}{\partial \eta}\right] \\ N_{s\theta}^{11} &= N_{\vartheta s}^{11} = \frac{E\delta}{1+\nu}\frac{1}{\rho}\frac{1}{2}\left[\frac{1}{2}\eta\frac{\partial U}{\partial \eta} + \frac{\partial V}{\partial \vartheta}\sec\alpha\right] \\ M_s^{11} &= E\delta L\tan^2\alpha\frac{1}{\rho^2}\left[\eta^2\frac{\partial^2 W}{\partial \eta^2} + \eta\frac{\partial W}{\partial \eta}\right] \\ M_\vartheta^{11} &= \nu M_s^{11} \\ M_{s\vartheta}^{11} &= M_{\theta s}^{11} = \frac{2E\delta}{1+\nu}L\tan^2\alpha\frac{1}{\rho^3}\eta\frac{\partial^2 W}{\partial\eta\partial\theta}\sec\alpha \end{aligned} \tag{12.29a-f}$$

in which the relation

$$k = 4\left(1-\nu^2\right)\tan^2\alpha\frac{1}{\rho^4} \tag{12.30}$$

obtained from the expression (12.23) has been used.

Note that the normal stresses N_s^{11} and N_θ^{11} are of the same order as that of N_s^1 and N_θ^1. It can be shown, however, that N_θ^1 and N_s^{11}vanish automatically.

Combining the two sets of solutions and when only the terms of the lowest order of $1/\rho$ are retained, one has

$$\begin{array}{lll} u = u^1 & v = v^1 & w = w^1 + w^{11} \\ N_s = N_s^1 & N_\theta = N_\theta^{11} & N_{s\theta} = N_{\theta s} = N_{s\theta}^1 \\ M_s = M_s^{11} & M_\theta = M_\theta^{11} & M_{s\theta} = M_{\theta s} = M_{s\theta}^{11} \end{array} \tag{12.31a-i}$$

Similarly, the transverse and tangential shearing forces defined by (12.10e and 12.20) are

$$S_\theta = S_\theta^{11} \quad S_s = S_s^{11} \quad T_s = T_s^1 = N_{s\theta}^1 \tag{12.32a-c}$$

Thus the two parts of the solutions, membrane and bending, are coupled by the lateral deflection, w; otherwise, they would be separable.

In view of (12.31, 12.32 and 12.28), and when solutions (12.24 and 12.25) are used, the stresses and moments may be given in the following final explicit forms:

$$\begin{aligned}
N_s &= -2E\delta\tan\alpha\left[\frac{C_2}{m^2-2\left(1-\nu^2\right)}\frac{1}{y^2}+\frac{3C_4}{m^2-7+2\nu}\frac{1}{y^4}\right]\frac{\cos}{\sin}\frac{n\pi\theta}{\theta_1}\\
N_\theta &= E\delta y^{-1}\tan\alpha\left\{y^\rho\left[C_5\cos\left(\rho\ln y\right)+C_6\sin\left(\rho\ln y\right)\right]\right.\\
&\quad\left.+y^{-\rho}\left[C_7\cos\left(\rho\ln y\right)+C_8\sin\left(\rho\ln y\right)\right]\right\}\frac{\cos}{\sin}\frac{n\pi\theta}{\theta_1}\\
N_{s\theta} &= T_s = \left(\mp\right)E\delta\left[\frac{C_4 6\tan\alpha}{m\left(m^2-7+2\nu\right)}\frac{1}{y^4}\right]\frac{\sin}{\cos}\frac{n\pi\theta}{\theta_1}\\
M_s &= \frac{2E\delta}{\rho^2}Ly\tan^2\alpha\left\{y^\rho\left[-C_5\sin\left(\rho\ln y\right)+C_6\cos\left(\rho\ln y\right)\right]\right.\\
&\quad\left.+y^{-\rho}\left[C_7\sin\left(\rho\ln y\right)-C_8\cos\left(\rho\ln y\right)\right]\right\}\frac{\cos}{\sin}\frac{n\pi\theta}{\theta_1}\\
M_\theta &= \nu M_s\\
S_\theta &= \mp\frac{2E\delta}{\rho^2}m\left(2-\nu\right)\frac{1}{y}\tan^2\alpha\left\{y^\rho\left[-C_5\sin\left(\rho\ln y\right)+C_6\cos\left(\rho\ln y\right)\right]\right.\\
&\quad\left.+y^{-\rho}\left[C_7\sin\left(\rho\ln y\right)-C_8\cos\left(\rho\ln y\right)\right]\right\}\frac{\sin}{\cos}\frac{n\pi\theta}{\theta_1}\\
S_s &= \frac{E\delta}{\rho}\frac{1}{y}\tan^2\alpha\left\{y^\rho\left[\left(-C_5+C_6\right)\cos\left(\rho\ln y\right)-\left(C_5+C_6\right)\sin\left(\rho\ln y\right)\right]\right.\\
&\quad\left.+y^{-\rho}\left[\left(C_7+C_8\right)\cos\left(\rho\ln y\right)-\left(C_7+C_8\right)\sin\left(\rho\ln y\right)\right]\right\}\frac{\cos}{\sin}\frac{n\pi\theta}{\theta_1}
\end{aligned} \tag{12.33a-h}$$

and

$$\begin{aligned}
\frac{\partial w}{\partial s} &= \frac{\partial w^{11}}{\partial s}\\
&= \frac{1}{2L}\rho\frac{1}{y^3}\left\{y^\rho\left[\left(C_5+C_6\right)\cos\left(\rho\ln y\right)-\left(C_5-C_6\right)\sin\left(\rho\ln y\right)\right]\right.\\
&\quad\left.-y^{-\rho}\left[\left(C_7-C_8\right)\cos\left(\rho\ln y\right)+\left(C_7+C_8\right)\sin\left(\rho\ln y\right)\right]\right\}\frac{\cos}{\sin}\frac{n\pi\theta}{\theta_1}
\end{aligned}$$

12.3 Particular Solutions due to Lateral Normal Loads

Let the lateral load given by (12.11) be confined to the form

$$p_{rn}\left(y\right) = p_n L^\beta y^{2\beta} \tag{12.34}$$

where p_n and β are prescribed constants.

One may assume a set of particular solutions in the similar form as given by expressions (12.9) except λ_n. In this case, λ_n is replaced by

$$\lambda^* = 2\beta + 3 \tag{12.35}$$

which is a known number. By solving the three algebraic (12.12) simultaneously, the particular solutions are readily obtained, provided that λ^* is not one of the roots of the determinant. However, when the load is uniformly distributed along meridians, $\beta = 0$ and $\lambda^* = 3$, which is one of the roots in the asymptotic case. In such cases, the approach needs to be modified. Since this is one of the most common loadings, the particular solution for this kind of uniform load is pursued below.

Because, in this case, λ^* is a finite constant as the parameter k approaches zero, the corresponding particular solution may be obtained from the equations of membrane theory of the system.

Letting $k = 0$ and having the independent variable s transformed to y, (12.7) are reduced to the following equilibrium equations of membrane theory:

$$\begin{aligned}
&\frac{1-\nu}{8}\left[y^2\frac{\partial^2 u}{\partial y^2} + 3y\frac{\partial u}{\partial y} - 8u\right] + \frac{1+\nu}{4}y\frac{\partial^2 u}{\partial y \partial \theta}\sec\alpha + \frac{\partial^2 u}{\partial \theta^2}\sec^2\alpha \\
&\quad + (2-\nu)\frac{\partial v}{\partial \theta}\sec\alpha + \frac{\partial w}{\partial \theta}\sec\alpha\tan\alpha = 0 \\
&\frac{1+\nu}{4}y\frac{\partial^2 u}{\partial y \partial \theta}\sec\alpha - \frac{3}{2}(1-\nu)\frac{\partial u}{\partial \theta}\sec\alpha + \frac{1}{4}y^2\frac{\partial^2 v}{\partial y^2} + \frac{3}{4}y\frac{\partial v}{\partial y} \\
&\quad + \frac{1-\nu}{2}\frac{\partial^2 v}{\partial \theta^2}\sec^2\alpha - (1-\nu)v + \frac{1}{2}\nu y\frac{\partial w}{\partial y}\tan\alpha - (1-\nu)w\tan\alpha = 0 \\
&\frac{\partial u}{\partial \theta}\sec\alpha + \frac{1}{2}\nu y\frac{\partial v}{\partial y} + v + w\tan\alpha = \frac{1-\nu^2}{E\delta}Lp_r y^2\frac{1}{\tan\alpha}
\end{aligned} \tag{12.36a-c}$$

where

$$p_r = p_n\frac{\cos\ n\pi\theta}{\sin\ \theta_1} \tag{12.37}$$

Let the particular solutions of (12.36) be assumed as below:

$$\begin{aligned}
u^p &= \mp(d_1 + d_2\ln y)\, y^2\frac{\sin\ n\pi\theta}{\cos\ \theta_1} \\
v^p &= (b_1 + b_2\ln y)\, y^2\frac{\cos\ n\pi\theta}{\sin\ \theta_1} \\
w^p &= e_1(1 + \ln y)\, y^2\frac{\cos\ n\pi\theta}{\sin\ \theta_1}
\end{aligned} \tag{12.38a-c}$$

in which d_1, d_2, b_1, b_2, and e_1 are constants to be determined. When these assumed solutions are substituted into (12.36) and after the sinusoidal functions and y^2 are cancelled, one will have three equations:

$$H_i + J_i \ln y = \frac{Lp_n}{\tan\alpha}\frac{1-\nu^2}{E\delta}\delta_{ir} \tag{12.39}$$

where J_i and H_i are the physical expressions and to-be-determined constants; δ_{ir} is Kronecker delta; and $i = s$, θ, and r.

By making the coefficients of both sides of (12.39) equal, there are two sets of algebraic equations, each containing three equations:

$$J_i = 0 \tag{12.40}$$

and

$$H_i = \frac{Lp_n}{\tan\alpha}\frac{1-\nu^2}{E\delta}\delta_{ir} \tag{12.41}$$

Only two of (12.40) are independent, because $\lambda^* = 3$ is one of the roots of the determinant. Thus, the five constants may be determined by the five independent (12.40 and 12.41). This results in

$$\begin{aligned} u^p &= \mp\frac{p_n}{\tan\alpha}\frac{L}{E\delta}\frac{m}{3}\left\{\frac{1}{2m^2}\left[2m^4 - 3(5-\nu)m^2 - 3(1+\nu)\right.\right. \\ &\quad \left.\left. + \left(m^2 - 7 + 2\nu\right)\ln y\right]\right\} y^2 \frac{\sin}{\cos}\frac{n\pi\theta}{\theta_1} \\ v^p &= \frac{p_n}{\tan\alpha}\frac{L}{E\delta}\frac{1}{6}\left[3(1-2\nu) - m^2\right] y^2 \frac{\cos}{\sin}\frac{n\pi\theta}{\theta_1} \\ w^p &= \frac{p_n}{\tan^2\alpha}\frac{L}{E\delta}\frac{1}{3}m^2\left(m^2 - 7 + 2\nu\right)(1 + \ln y)\frac{\cos}{\sin}\frac{n\pi\theta}{\theta_1} \end{aligned} \tag{12.42a-c}$$

When these displacements are substituted into the expressions (12.4a-c) with $k = 0$, the corresponding stress resultants are

$$\begin{aligned} N_s^p &= \frac{p_n L}{\tan\alpha}\frac{1}{6}\left(3 - m^2\right) y^2 \frac{\cos}{\sin}\frac{n\pi\theta}{\theta_1} \\ N_\theta^p &= \frac{p_n L}{\tan\alpha} y^2 \frac{\cos}{\sin}\frac{n\pi\theta}{\theta_1} \\ N_{s\theta}^p &= \pm\frac{p_n L}{\tan\alpha}\frac{m}{3} y^2 \frac{\sin}{\cos}\frac{n\pi\theta}{\theta_1} \end{aligned} \tag{12.43a-c}$$

These particular solutions combined with those given by solutions (12.24, 12.25 and 12.33) constitute the complete solutions.

12.4 An Example – A Semi-Circular Cone Subjected to Normal Loads

For the purpose of illustration, take a truncated semi-circular cone with the two generators simply supported as an example. Thus, the lower set of the sinusoidal solutions is to be used. Let it be clamped at the smaller end at $s = L_1$ and free at the other end where $s = L$, so that

$$u = v = w = \frac{\partial w}{\partial s} = 0 \quad \text{at } y = \sqrt{\frac{L_1}{L}} \tag{12.44a-d}$$

$$N_s = T_s = M_s = S_s = 0 \quad \text{at } y = 1 \tag{12.44e-h}$$

By using boundary conditions of (12.44a,b,e,f), constants C_1, C_2, C_3, and C_4 can be determined; then, the other four constants can be determined by the remaining four boundary conditions. Hence, instead of eight simultaneous equations, one needs to solve two sets of four simultaneous equations.

For numerical computations, the following values are assumed

$$\alpha = 75^\circ \quad \nu = \frac{1}{3} \quad \sqrt{\frac{L_1}{L}} = 0.90 \tag{12.45a-c}$$

Considering t/R as a parameter where R is the principal radius at a section of thickness t, one obtains $\delta = (t/R)\cos\alpha$. From (12.16 and 12.17) and (12.21 and 12.22) for asymptotic values, the eight roots of λ are listed in Table 12.1, for $n = 1$ and 2.

Table 12.1. The values of λ

λ	$\frac{t}{R}$	$n = 1$	$n = 2$	Asymptotic Values
$\lambda_{\substack{1\\2}}$	0.004	±0.999999	±1.0523	±1
	0.006	±0.999997	±1.1142	±1
	0.008	±0.999995	±1.1955	+1
$\lambda_{\substack{3\\14}}$	0.004	±3.00003	±2.6851	±3
	0.006	±3.00007	±2.9663	±3
	0.008	+3.00013	±2.9397	±3
$\lambda_{\substack{57\\68}}$	0.004	$\pm 153.27(1.0027 \pm i)$	$\pm 152.75(1.0099 \pm i)$	$\pm 153.48(1 \pm i)$
	0.006	$\pm 125.09(1.0053 \pm i)$	$\pm 124.51(1.0149 \pm i)$	$\pm 125.32(1 \pm i)$
	0.008	$\pm 108.28(1.0045 \pm i)$	$\pm 107.77(1.0198 \pm i)$	$\pm 108.53(1 \pm i)$

The asymptotic solutions of displacements, stresses, and moments computed may be given in the form

$$G_n(y, \theta) = g_n(y) \frac{\sin n\pi\theta}{\cos \ \theta_1} \qquad n = 1 \text{ and } 2 \tag{12.46}$$

in which the function $g_n(y)$ are presented in Figs. 12.1 to 12.7.

When $n = 1$, the solution represents the response to a symmetric lateral load as shown in Fig. 12.8(a). By a proper combination of solutions for

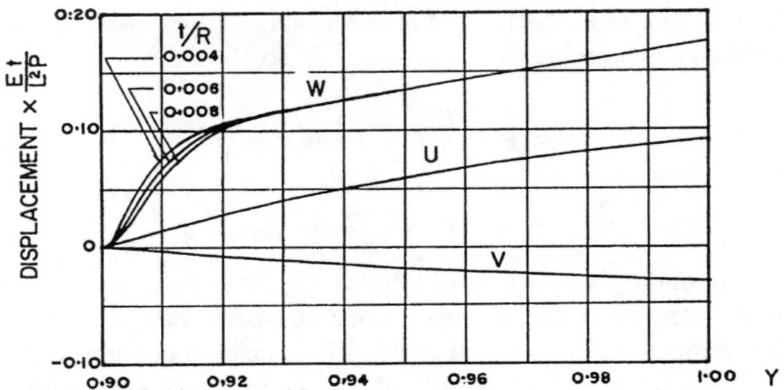

Fig. 12.1. Displacements u, v and w $(n = 1)$

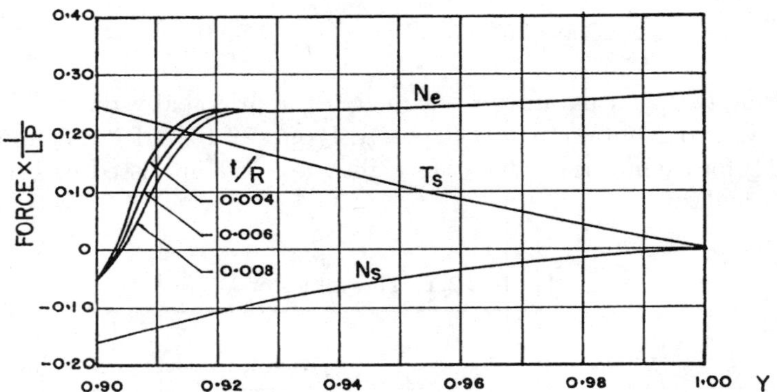

Fig. 12.2. Membrane forces N_θ,T_s and N_s $(n = 1)$

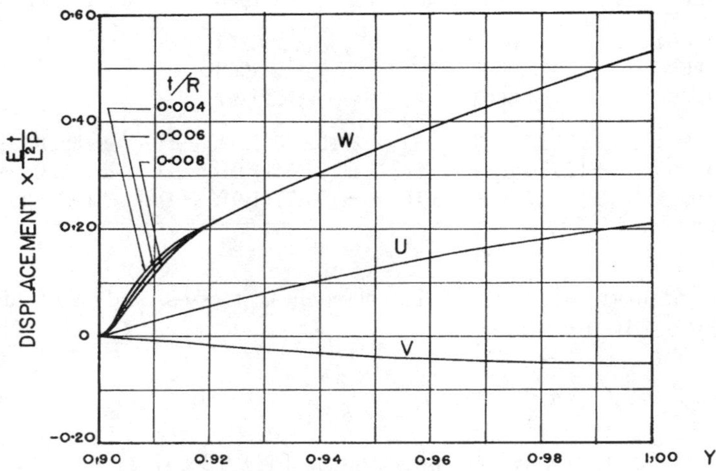

Fig. 12.3. Displacements u, v and w $(n = 2)$

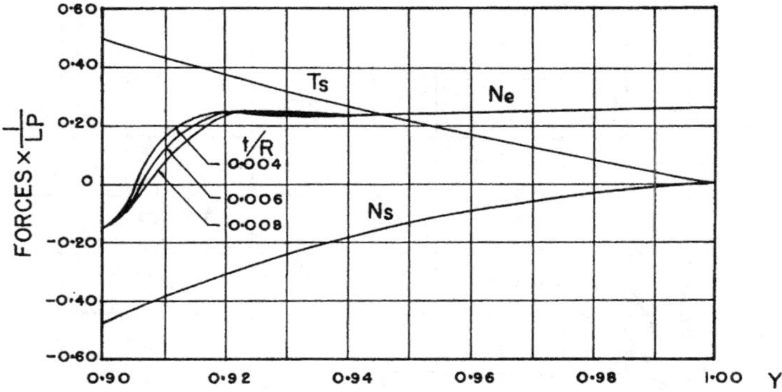

Fig. 12.4. Membrane forces N_θ, T_s and N_s ($n = 2$)

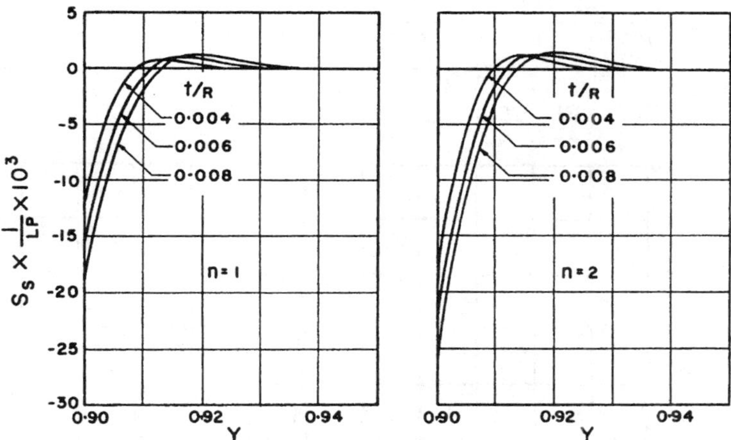

Fig. 12.5. Transverse shearing forces S_s

$n = 1$ and 2, one may have the response to an asymmetric load as shown in Fig. 12.8(b).

The present asymptotic solutions are exact and applicable to conical shells when

$$\left[\frac{1}{12}\left(\frac{t}{R}\cos\alpha\right)^2\right]^{1/2} \ll 1 \tag{12.47}$$

When the above parameter is very small (as were those given in the example), the present explicit solutions will provide a good approximation for conical shells of constant thickness, which will be studied in the next chapter.

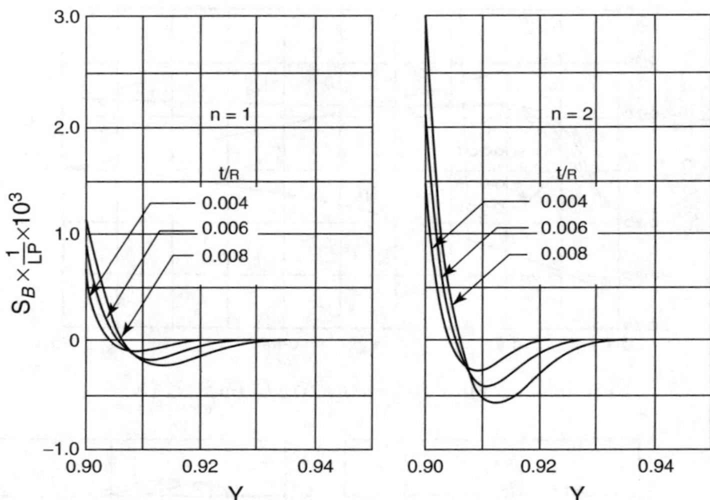

Fig. 12.6. Transverse shearing force S_θ

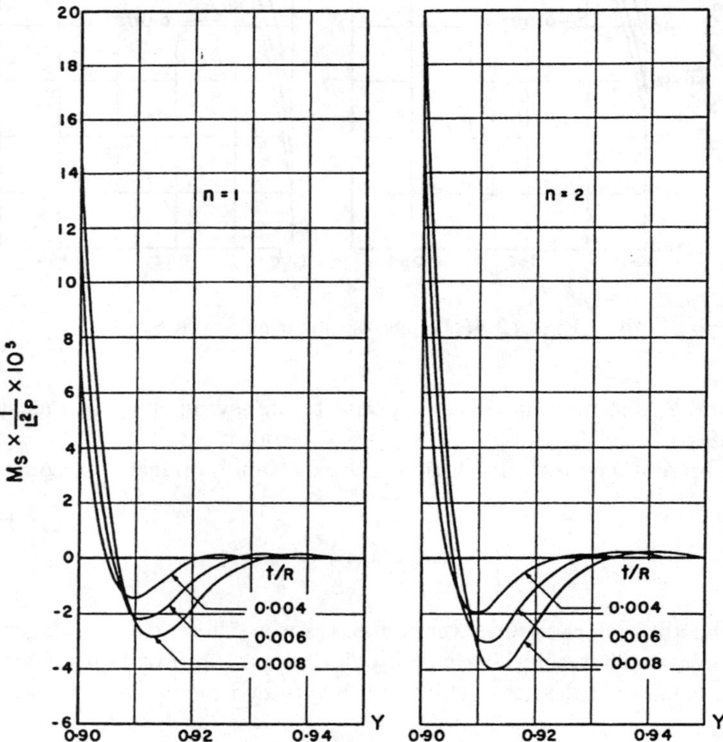

Fig. 12.7. Normal moment M_s

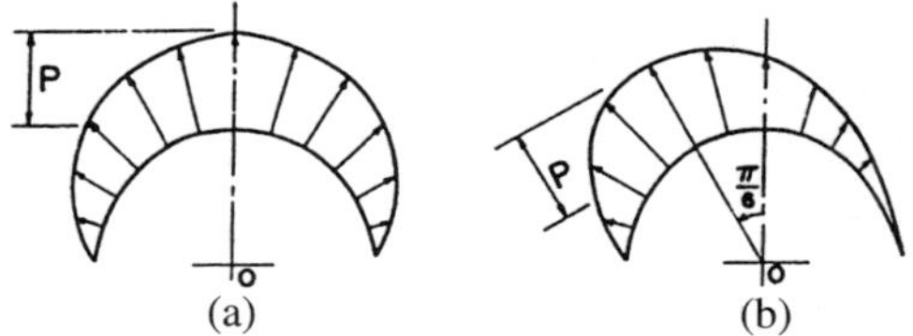

Fig. 12.8. (**a**) Symmetric load (**b**) Asymmetric load

Exercises

1. Verify (12.16 and 12.17).
2. Show that N_θ^1 and Ns^{11} vanish automatically.
3. Verify solutions (12.42) by substituting (12.38) into (12.36).
4. Determine the eight λ's for the cone defined by (12.45) for $t/R = 0.01$ for both $n = 1$ and 2.

13

Asymptotic Solutions of Conical Shells of Constant Thickness

The solutions of conical shells with linearly varying thickness presented in the last chapter have suggested a new approach to obtain asymptotic solutions of conical shells with constant thickness. One may assume that the two types of conical shells would behave alike if the ratios of thickness to the radius of the smaller end of the cone is very small. Under this assumption, asymptotic solutions of conical shells with constant thickness are obtained in this chapter. Two numerical examples are given for comparison with the present solutions: the semi-circular conical segment discussed in the last chapter and a complete cone. A solution of the latter is given in [12].

13.1 Basic Equations

The set of relations between stress resultants and displacements for conical shells of (12.4) presented in the last chapter may be used for the present study with the following two changes: (1) The constants D and k defined by (12.5a) and (12.6) in the last chapter are now changed to D and k defined in (13.2) in this chapter. (2) The meridional coordinate, s, is changed to y as defined by (12.8) which reads $y = (s/L)^{1/2}$ in the last chapter.

It has been learned from the results of the last chapter that for thin shells: (1) The bending has no effects to the plane stress resultants; therefore, the terms associated with k may be dropped for these stresses. (2). The circumferential displacement, u, and meridional displacement, v, have no effect to the moments. Thus the terms associated with these displacements in moments can be discarded. With these simplifications, using the same nomenclature, the following relations are obtained from (12.4) in the last chapter:

$$N_s = \frac{D}{L}\frac{1}{y^2}\left[\frac{1}{2}yv^{\cdot} + \nu\left(u'\sec\alpha + v + w\tan\alpha\right)\right]$$

$$N_\theta = \frac{D}{L}\frac{1}{y^2}\left[u'\sec\alpha + v + w\tan\alpha + \frac{\nu}{2}yv^{\cdot}\right]$$

$$
\begin{aligned}
N_{s\theta} &= N_{\theta s} = \frac{D}{L}\frac{1-\nu}{2y^2}\left[\frac{1}{2}yu^{\cdot} - u + v'\sec\alpha\right] \\
M_s &= \frac{Dk}{y^4}\left[\frac{1}{4}\left(y^2w^{\cdot\cdot} - yw^{\cdot}\right) + \nu\left(w''\sec^2\alpha + \frac{1}{2}yw^{\cdot}\right)\right] \\
M_\theta &= \frac{Dk}{y^4}\left[w''\sec^2\alpha + \frac{1}{2}yw^{\cdot} + \frac{\nu}{4}\left(y^2w^{\cdot\cdot} - yw^{\cdot}\right)\right] \\
M_{s\theta} &= M_{\theta s} = \frac{Dk\,(1-\nu)}{y^4}\left[\frac{1}{2}yw'^{\cdot} - w'\right]\sec\alpha
\end{aligned}
\qquad (13.1\text{a-f})
$$

The D and k are redefined below:

$$D = \frac{Et}{1-\nu^2} \qquad (13.2\text{a})$$

$$k = \frac{1}{12}\left(\frac{t}{L}\right)^2 = \frac{1}{12}\left(\frac{t}{R}\right)^2\sin^2\alpha \qquad (13.2\text{b})$$

where E is Young's modulus of elasticity; ν is Poisson's ratio; t is the thickness; L is the slanted length from the apex to the larger end, L_1 to the smaller end of the shell; and $R = L\ \sin\alpha$ is the radius of the shell at $y = 1$.

When the equilibrium equation of moments about the normal of a surface element is overlooked, the other five equilibrium equations derived from (12.la-e) in the last chapter are

$$
\begin{aligned}
& yN_s^{\cdot}/2 + N_s + N'_{\theta s}\sec\alpha - N_\theta = -P_sLy^2 \\
& yN_{s\theta}^{\cdot}/2 + 2N_{s\theta} + N'_\theta\sec\alpha - Q_\theta\tan\alpha = -P_\theta Ly^2 \\
& yQ_s^{\cdot}/2 + Q_s + Q'_\theta\sec\alpha + N_\theta\tan\alpha = P_rLy^2 \\
& yM_s^{\cdot}/2 + M_s + M'_{\theta s}\sec\alpha - M_\theta = Ly^2Q_s \\
& yM_{s\theta}^{\cdot}/2 + 2M_{s\theta} + M'_\theta\sec\alpha = Ly^2Q_\theta
\end{aligned}
\qquad (13.3\text{a-e})
$$

where Q_s and Q_θ are the transverse shear forces per unit length acting on sections perpendicular to the s and θ directions; and P_r, P_s, and P_θ are surface loads per unit area in normal, meridional, and circumferential directions, respectively.

The eleven equations in (13.1) and (13.3) govern the eleven unknowns involved. When (13.3d and 13.3e) are used to eliminate the transverse shearing forces Q_s and Q_θ in (13.3b and 13.3c), the first three (13.3a,b,c) become:

$$
\begin{aligned}
& yN_s^{\cdot}/2 + N_s + N'_{\theta s}\sec\alpha - N_\theta = -P_sLy^2 \\
& yN_{s\theta}^{\cdot}/2 + 2N_{s\theta} + N'_\theta\sec\alpha - \left(1/Ly^2\right)\left[yM_{s\theta}^{\cdot}/2\right. \\
& \quad \left. + \ 2M_{s\theta} + M'_\theta\sec\alpha\right]\tan\alpha = -P_\theta Ly^2 \\
& N_\theta\tan\alpha + \left(1/Ly^2\right)\left[\frac{1}{4}y^2M_s^{\cdot\cdot} + \frac{3}{4}yM_s^{\cdot} + \left(yM_{s\theta}'^{\cdot} + 2M'_{s\theta}\right)\sec\alpha\right. \\
& \quad \left. + M''_\theta\sec^2\alpha - yM_\theta^{\cdot}/2\right] = P_rLy^2
\end{aligned}
\qquad (13.4\text{a-c})
$$

In the following discussion, a truncated conical shell is considered. One of the following three types of edge conditions usually exists, and each type has four boundary conditions to be satisfied at the two circular ends: built-in edges:

$$u = 0 \quad v = 0 \quad w = 0 \quad \text{and} \quad w^{\cdot} = 0 \tag{13.5a-d}$$

free edges:

$$N_s = 0 \quad M_s = 0 \quad S_s = 0 \quad \text{and} \quad T_s = 0 \tag{13.5e-h}$$

simply supported edges

$$w = 0 \quad M_s = 0 \quad N_s = 0 \quad \text{or} \quad v = 0 \quad \text{and} \quad T_s = 0 \quad \text{or} \quad u = 0 \tag{13.5i-n}$$

where

$$S_s = Q_s + \left(1/Ly^2\right) M'_{s\theta} \sec\alpha \tag{13.6a}$$

$$T_s = N_{s\theta} - \left(1/Ly^2\right) M_{s\theta} \tan\alpha \tag{13.6b}$$

A special type of boundary conditions will be discussed in Sect. 13.5.2.

The frustum may be a complete cone or a segment of a cone. For the latter, the two generator edges are considered to be simply supported so that

$$w = 0 \quad v = 0 \quad N_\theta = 0 \quad \text{and} \quad M_\theta = 0 \tag{13.7a-d}$$

at $\theta = 0$ and θ_1, where $\theta_1 < 2\pi$, is the angle between the two generator edges. The normal reactions along those two edges are given by

$$S_\theta = Q_\theta + {}^1\!/_2\ (y/L)\, M^{\cdot}_{\theta s} \tag{13.8}$$

Substituting (13.1) into (13.4), three equations for the three unknown displacement components are obtained. Instead of dealing with these three components, each component is divided into three parts: membrane, bending effect and particular solutions due to lateral loads, if any. Denoting these three parts by superscripts I, II, and P, respectively, the three displacement components may be expressed as

$$u = u^{\mathrm{I}} + u^{\mathrm{II}} + u^{\mathrm{P}} \qquad v = v^{\mathrm{I}} + v^{\mathrm{II}} + u^{\mathrm{P}} \tag{13.9a,b}$$

$$w = w^{\mathrm{I}} + w^{\mathrm{II}} + w^{\mathrm{P}} \tag{13.9c}$$

These three parts of the solutions will be analyzed separately in the following sections.

13.2 Membrane Solutions

The membrane solutions of conical shells with constant thickness are well known. However, available solutions are presented in the form of stresses. In the following discussion, the displacement components will be presented in explicit form.

When (13.1) with $k = 0$ are substituted into (13.4), the three equilibrium equations become

$$\begin{aligned}
&\frac{1-\nu}{8}\left(y^2 u^{\cdot\cdot} + y u^{\cdot} - 4u\right) + \left(u'' \sec\alpha + \frac{1+\nu}{4} y v'^{\cdot}\right. \\
&\quad \left. + \frac{3-\nu}{2} v' + w' \tan\alpha\right) \sec\alpha = -\frac{L^2 y^4}{D} P_\theta \\
&\left(\frac{1+\nu}{4} y u'^{\cdot} - \frac{3-\nu}{2} u' + \frac{1-\nu}{2} v'' \sec\alpha\right) \sec\alpha + \frac{1}{4}\left(y^2 v^{\cdot\cdot} + y v^{\cdot}\right) \\
&\quad -v + \left(\frac{\nu}{2} y w^{\cdot} - w\right) \tan\alpha = -\frac{L^2 y^4}{D} P_s \\
&\left(u' \sec\alpha + v + w \tan\alpha + \frac{\nu}{2} y\, v^{\cdot}\right) \tan\alpha = \frac{L^2 y^4}{D} P_r
\end{aligned} \tag{13.10a-c}$$

Let

$$u = A y^{\lambda} \frac{\sin\, n\pi\theta}{\cos\ \theta_1} \qquad v = B y^{\lambda} \frac{\cos\, n\pi\theta}{\sin\ \theta_1} \qquad w = C y^{\lambda} \frac{\cos\, n\pi\theta}{\sin\ \theta_1} \tag{13.11a-c}$$

where λ is an unknown constant. Substituting (13.11) into the homogeneous part of (13.10) and eliminating the y and sinusoidal functions, one has three homogeneous algebraic equations for the three unknown constants A, B and C. Letting the determinant of the equations vanish in order to have nontrivial solution results in the following characteristic equation for λ:

$$\lambda^2 \left(\lambda^2 - 4\right) = 0 \tag{13.12}$$

When the values of λ are determined and substituted back into the algebraic equations, one may express the constants A and B in terms of C. Then the first part of the solutions of the displacement components are expressed in the following form:

$$\begin{aligned}
u^1 &= \mp\frac{1}{m}\left\{\frac{m^2}{m^2-1}\left[C_1 - C_2\left(\frac{1-\nu}{2\left(m^2-1\right)} - \ln y\right)\right]\right. \\
&\quad \left. + C_3\, y^2 + \frac{C_4}{y^2}\frac{m^2 - 2\left(1+\nu\right)}{m^2-4}\right\} \frac{\sin\, n\pi\theta}{\cos\ \theta_1} \\
v^1 &= \left\{\frac{1}{m^2-1}\left[C_1 - C_2\left(\frac{m^2-\nu}{2\left(m^2-1\right)} - \ln y\right)\right]\right. \\
&\quad \left. + \frac{C_4}{y^2}\frac{2}{m^2-4}\right\} \frac{\cos\, n\pi\theta}{\sin\ \theta_1} \\
w^1 &= \frac{1}{\tan\alpha}\left[C_1 + C_2 \ln y + C_3 y^2 + \frac{C_4}{y^2}\right] \frac{\cos\, n\pi\theta}{\sin\ \theta_1}
\end{aligned} \tag{13.13a-c}$$

where $m = (n\pi/\theta_1)\sec\alpha$. The corresponding stress resultants may be obtained from (13.1) as

$$N_s^1 = \frac{Et}{L}\left[\frac{C_2}{y^2}\frac{1}{2(m^2-1)} - \frac{C_4}{y^4}\frac{2}{m^2-4}\right]\frac{\cos}{\sin}\frac{n\pi\theta}{\theta_1}$$
$$N_{s\theta}^1 = \mp\frac{Et}{L}\left[\frac{C_4}{y^4}\frac{2}{m(m^2-4)}\right]\frac{\sin}{\cos}\frac{n\pi\theta}{\theta_1} \qquad (13.14\text{a-b})$$

N_θ^1 vanishes automatically.

13.3 Solutions of Bending Effect

It is learned from the last chapter that the displacements due to bending may be given in the following form

$$u^{\mathrm{II}} = U/\gamma^2 \quad v^{\mathrm{II}} = V/\gamma \quad w^{\mathrm{II}} = W \qquad (13.15\text{a-c})$$

where

$$\gamma^4 = 16/k \qquad (13.16)$$

Note that $1/\gamma$ is of the order $(t/R)^{1/2}$. Furthermore, the y functions of U, V and W may be expressed in terms of $y^{c\gamma}$ where c is a finite constant. Thus the derivatives with respect to y will change the orders of the magnitude of the functions concerned. In order to avoid this, a new variable η is introduced such that

$$\eta = y^\gamma \qquad (13.17)$$

When expressions (13.15 and 13.17) are substituted into the elastic relations (13.1), retaining only the terms of the lowest order of $1/\gamma$ yields

$$N_s^{\mathrm{II}} = \frac{D}{L}\left[\frac{1}{2}\eta\frac{\partial V}{\partial\eta} + \nu W\tan\alpha\right]\eta^{-(2/\gamma)}$$
$$N_\theta^{\mathrm{II}} = \frac{D}{L}\left[W\tan\alpha + \frac{1}{2}\nu\eta\frac{\partial V}{\partial\eta}\right]\eta^{-(2/\gamma)}$$
$$N_{s\theta}^{\mathrm{II}} = N_{\theta s}^{\mathrm{II}} = \frac{D}{L}\frac{1-\nu}{2}\frac{1}{\gamma}\left[\frac{\eta}{2}\frac{\partial U}{\partial\eta} + \frac{\partial V}{\partial\theta}\sec\alpha\right]\eta^{-(2/\gamma)} \qquad (13.18\text{a-g})$$
$$M_s^{\mathrm{II}} = D\frac{4}{\gamma}\left[\eta^2\frac{\partial^2 W}{\partial\eta^2} + \eta\frac{\partial W}{\partial\eta}\right]\eta^{-(4/\gamma)}$$
$$M_\theta^{\mathrm{II}} = \nu M_s^{\mathrm{II}}$$
$$M_{s\theta}^{\mathrm{II}} = M_{\theta s}^{\mathrm{II}} = D\frac{8}{\gamma^3}(1-\nu)\eta\frac{\partial^2 W}{\partial\eta\partial\theta}\sec\alpha\eta^{-4/\gamma}$$

It is of interest to note that N_s^{II} and N_θ^{II} are of the same order of $(1/\gamma)^\circ$ as that in the membrane solutions.

Transforming the variable y to η and using the asymptotic relations (13.18), the homogenous part of (13.4) with only the terms of the lowest order of $1/\gamma$ retained results in the following three equations:

$$\begin{aligned}
&\eta^2\frac{\partial^2 V}{\partial\eta^2}+\eta\frac{\partial V}{\partial\eta}+2\nu\eta\frac{\partial W}{\partial\eta}\tan\alpha=0\\
&\frac{1}{4}\frac{1-\nu}{2}\left[\eta^2\frac{\partial^2 U}{\partial\eta^2}+\eta\frac{\partial U}{\partial\eta}+2\eta\frac{\partial^2 V}{\partial\theta\partial\eta}\sec\alpha\right]\\
&\quad+\left[\frac{\partial W}{\partial\theta}\tan\alpha+\frac{1}{2}\nu\eta\frac{\partial^2 V}{\partial\theta\partial\eta}\right]\sec\alpha=0\\
&\eta^4\frac{\partial^4 W}{\partial\eta^4}+6\eta^3\frac{\partial^3 W}{\partial\eta^3}+7\eta^2\frac{\partial^2 W}{\partial\eta^2}+\eta\frac{\partial W}{\partial\eta}\\
&\quad+\eta^{4/\gamma}\left[W\tan^2\alpha+\frac{1}{2}\nu\eta\frac{\partial V}{\partial\eta}\tan\alpha\right]=0
\end{aligned}\tag{13.19a-c}$$

Integrating (13.19a) with respect to η assumes

$$\eta(\partial V/\partial\eta)=-2\nu W\tan\alpha \tag{13.20}$$

in which, without loss of generality, an integration constant has been dropped. Substituting (13.20) into (13.19c) yields

$$\eta^4\frac{\partial^4 W}{\partial\eta^4}+6\eta^3\frac{\partial^3 W}{\partial\eta^3}+7\eta^2\frac{\partial^2 W}{\partial\eta^2}+\eta\frac{\partial W}{\partial\eta}+\eta^{4/\gamma}\left(1-\nu^2\right)W\tan^2\alpha=0 \tag{13.21}$$

In the last equation, the factor $\eta^{4/\gamma}$ is an obstacle to integration. This factor, however, may be expanded in power series of η. In order to have an orthogonal set of series, it is expanded in Legendre polynomials. When this is done (see Appendix at the end of this chapter) and the term of the lowest order of $1/\gamma$ is retained

$$\eta^{4/\gamma}=\frac{\gamma}{\gamma+4} \tag{13.22}$$

Substituting (13.22) into (13.21) and assuming

$$W=(1/\tan\alpha)\,\bar{C}\eta^{\lambda}\frac{\cos n\pi\theta}{\sin\ \theta_1} \tag{13.23}$$

Equation (13.21) results in a characteristic equation:

$$\lambda^4+\frac{\gamma}{\gamma+4}\left(1-\nu^2\right)\tan^2\alpha=0 \tag{13.24}$$

which gives

$$\lambda=\pm q\,(1\pm i) \tag{13.25}$$

where

$$q=\left|{}^1\!/\!_4\left[\gamma/\left(\gamma+4\right)\right]\left(1-\nu^2\right)\tan^2\alpha\right|^{1/4} \tag{13.26}$$

Letting

$$V = \bar{B}\eta^{\lambda} \frac{\cos\ n\pi\theta}{\sin\ \theta_1} \quad U = \bar{A}\eta^{\lambda} \frac{\sin\ n\pi\theta}{\cos\ \theta_1} \tag{13.27a,b}$$

and using (13.20 and 13.19b), one has

$$\bar{B} = -\frac{2\nu}{\lambda}\bar{C} \quad \bar{A} = \pm\frac{4}{\lambda^2}(2+\nu)\,m\bar{C} \tag{13.28a,b}$$

where $\bar{A}, \bar{B}$ and $\bar{C}$ are complex numbers. When the complex numbers are transformed into real numbers and the solutions are expressed in the variable y and denoting

$$\gamma q = \rho \tag{13.29}$$

the solutions given in (13.15) and the induced stress resultants and moments obtained from (13.6, 13.8 and 13.18) assume the following final form:

$$\begin{aligned}
w^{\mathrm{II}} &= \frac{1}{\tan\alpha}\{y^{\rho}\left[C_5\cos(\rho\ln y) + C_6\sin(\rho\ln y)\right] \\
&\quad + y^{-\rho}\left[C_7\cos(\rho\ln y) + C_8\sin(\rho\ln y)\right]\}\frac{\cos\ n\pi\theta}{\sin\ \theta_1} \\
v^{\mathrm{II}} &= \frac{\nu}{\rho}Y_1\frac{\cos\ n\pi\theta}{\sin\ \theta_1} \\
u^{\mathrm{II}} &= \mp\left[2(2+\nu)\frac{m}{\rho^2}\right]Y_2\frac{\cos\ n\pi\theta}{\sin\ \theta_1} \\
N_S^{\mathrm{II}} &= 0 \\
N_\theta^{\mathrm{II}} &= \frac{Et}{L}w^{\mathrm{II}}\tan\alpha\frac{1}{y^2}\frac{\cos\ n\pi\theta}{\sin\ \theta_1} \\
M_s^{\mathrm{II}} &= \frac{2Et}{\rho^2}\tan\alpha\frac{1}{y^4}Y_2\frac{\cos\ n\pi\theta}{\sin\ \theta_1} \\
M_\theta^{\mathrm{II}} &= \nu M_s^{\mathrm{II}} \\
M_{s\theta}^{\mathrm{II}} &= \mp\frac{2(1-\nu)E}{\rho^3}\frac{m}{y^4}\tan\alpha\{y^{\rho}\left[(C_5+C_6)\cos(\rho\ln y)+(C_6-C_5)\sin(\rho\ln y)\right] \\
&\quad - y^{-\rho}\left[(C_7-C_8)\cos(\rho\ln y) + (C_7+C_8)\sin(\rho\ln y)\right]\}\frac{\cos\ n\pi\theta}{\sin\ \theta_1} \\
S_s^{\mathrm{II}} &= \frac{Et}{L\rho}\tan\alpha\frac{1}{y^6}Y_1\frac{\cos\ n\pi\theta}{\sin\ \theta_1} \\
S_\theta^{\mathrm{II}} &= \mp\frac{Et}{L\rho^2}\frac{2(2-\nu)}{y^6}m\tan\alpha Y_2\frac{\sin\ n\pi\theta}{\cos\ \theta_1}
\end{aligned} \tag{13.30a-h}$$

where

$$\begin{aligned}
Y_1 &= y^{\rho}\left[(C_6-C_5)\cos(\rho\ln y) - (C_5+C_6)\sin(\rho\ln y)\right] \\
&\quad + y^{-\rho}\left[(C_7+C_8)\cos(\rho\ln y) - (C_7-C_8)\sin(\rho\ln y)\right] \\
Y_2 &= y^{\rho}\left[C_6\cos(\rho\ln y) - C_5\sin(\rho\ln y)\right] - y^{-\rho}\left[C_8\cos(\rho\ln y) - C_7\sin(\rho\ln y)\right]
\end{aligned} \tag{13.30i,j}$$

13.4 A Particular Solution

Consider a conical shell subjected to an outward lateral normal load which is constant along the meridian and has a sinusoidal distribution in the circumferential direction. This is the case treated in the last chapter for the linearly varying thickness of conical shells. The membrane solutions of (13.10) are used for the particular solution.

Let

$$P_\theta = P_s = 0 \quad P_r = p_n \frac{\cos\, n\pi\theta}{\sin\ \theta_1} \tag{13.31a-c}$$

and assume

$$u^p = d_1 y^4 \frac{\sin\, n\pi\theta}{\cos\ \theta_1} \quad v^p = d_2 y^4 \frac{\cos\, n\pi\theta}{\sin\ \theta_1} \quad w^p = d_3 y^4 \frac{\cos\, n\pi\theta}{\sin\ \theta_1} \tag{13.32a-c}$$

where d_1, d_2, and d_3 are coefficients to be determined by substituting (13.31 and 13.32) into (13.10). This results in

$$\begin{aligned} u^p &= \mp \frac{p_n L^2}{Et} \frac{1}{12} \frac{m}{\tan\alpha} \left[11 + 2\nu - m^2\right] y^4 \frac{\sin\, n\pi\theta}{\cos\ \theta_1} \\ v^p &= \frac{p_n L^2}{Et} \frac{1}{12\tan\alpha} \left[3\left(1 - 2\nu\right) - m^2\right] y^4 \frac{\cos\, n\pi\theta}{\sin\ \theta_1} \\ w^p &= \frac{p_n L^2}{Et} \frac{1}{12\tan\alpha} \left(m^2 - 1\right)\left(m^2 - 9\right) y^4 \frac{\cos\, n\pi\theta}{\sin\ \theta_1} \end{aligned} \tag{13.33a-c}$$

The corresponding stresses obtained from (13.1) with $k = 0$ are

$$\begin{aligned} N_s^p &= \frac{p_n L}{6\tan\alpha} \left(3 - m^2\right) y^2 \frac{\cos\, n\pi\theta}{\sin\ \theta_1} \\ N_\theta^p &= \frac{p_n L}{\tan\alpha} y^2 \frac{\cos\, n\pi\theta}{\sin\ \theta_1} \\ N_{\theta s}^p &= \mp \frac{p_n L}{3\tan\alpha} m y^2 \frac{\sin\, n\pi\theta}{\cos\ \theta_1} \end{aligned} \tag{13.34a-c}$$

By retaining the solutions of the lowest order of $1/\rho$, one finally has the complete solutions for the shell subjected to lateral load of (13.31) as follows:

$$\begin{gathered} u = u^1 + u^p \qquad v = v^1 + v^p \qquad w = w^1 + w^{11} + w^p \\ N_s = N_s^1 + N_s^p \qquad N_\theta = N_\theta^{11} + N_\theta^p \qquad N_{s\theta} = N_{\theta s} = T_s = N_{s\theta}^1 + N_{s\theta}^p \\ M_s = M_s^{11} \qquad M_\theta = M_\theta^{11} \qquad M_{s\theta} = M_{\theta s} = M_{s\theta}^{11} \\ S_s = S_s^{11} \qquad S_\theta = S_\theta^{11} \end{gathered} \tag{13.35a-k}$$

It may be of interest to note that the vanishing of N_θ^{I} and N_s^{II} makes the two parts of the solutions, membrane (including the particular solutions in this case) and bending, coupled only in the normal displacement, w; otherwise these two parts could be separable. The eight constants involved $(C_{l}, \ldots, C_8)$ are to be determined by the boundary conditions at the two circular ends. One may notice that there are two and only two boundary conditions involving membrane solutions. Therefore, in the determination of the eight constants, one merely needs to solve two four-by-four matrices instead of one eight-by-eight matrix as in the last chapter.

13.5 Numerical Examples

For comparison, there are two examples given: One is for the semi-circular cone given in the last chapter; the other is for a complete cone studied by Clark and Garibotti [12].

13.5.1 The Semi-Circular Cone Studied in Chap. 12

The semi-circular conical shell segment, which was given in the last chapter for the conical shell with linearly varying thickness, is used in order to make a comparison for the two solutions.

The semi-circular segment cone has two simply supported generators with the smaller end fixed and the other end free. The lower set of sinusoidal functions is used with the boundary conditions (13.5a-d) at $y = (L_1/L)^{1/2}$ and conditions (13.5e-h) at $y = 1$.

The following material and geometrical constants are used:

$$\nu = 1/3 \qquad \alpha = 75^\circ \qquad \text{and} \qquad (L_1/L)^{1/2} = 0.90 \qquad (13.36\text{a-c})$$

Numerical results for $t/R = 0.006$ and $n = 1, 2$ were computed. The results are given in the form

$$F_n(y,\theta) = f_n(y) \frac{\sin n\pi\theta}{\cos\ \theta_1} \qquad n = 1 \text{ and } 2 \qquad (13.37)$$

The functions $f_n(y)$ are shown in solid lines in Figs. 13.1 to 13.6. The respective functions obtained in the study of conical shells of linearly varying thickness in the last chapter are also shown in these figures by dotted lines, if there are any differences.

13.5.2 A Cantilever Complete Cone

In this example, the complete conical cantilever frustum treated by Clark and Garibotti [12] is considered. The frustum is fixed at the smaller end. At the larger free end, a rigid plate is attached and a moment, M, is applied about the horizontal axis of the plate. Thus the solutions are symmetrical about the vertical axis through the center of the cone. For such a complete cone,

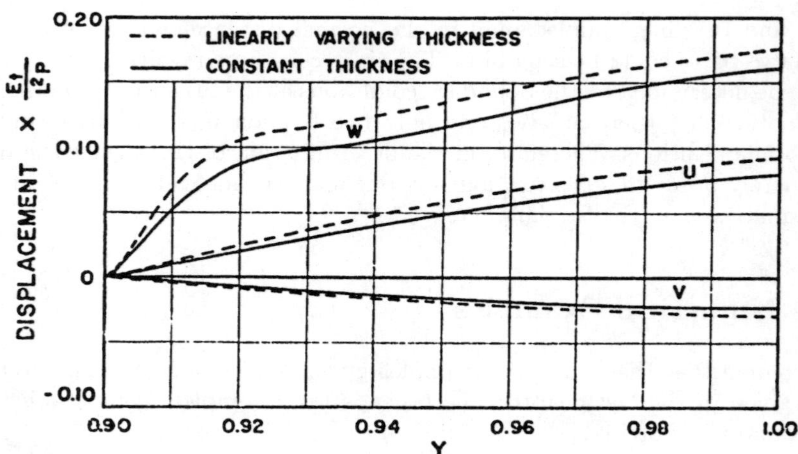

Fig. 13.1. Displacements u, v, and w $(n = 1)$

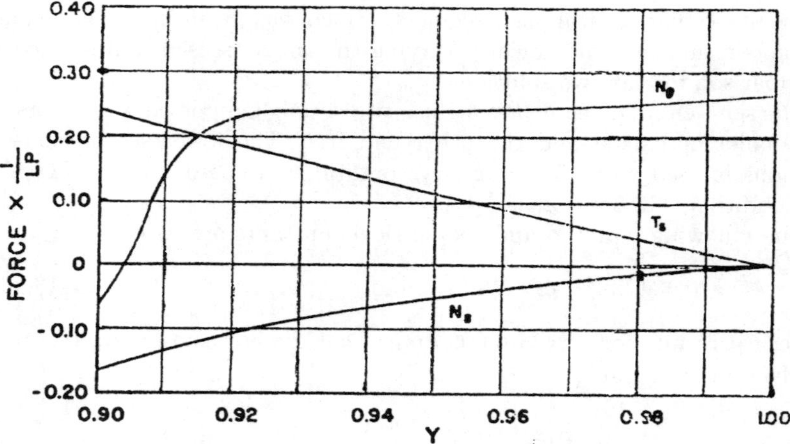

Fig. 13.2. Membrane forces N_θ, T_s, and N_s $(n = 1)$

the upper set of sinusoidal functions of the solutions is used with the angle θ measured from the vertical line taking $n = 1$, and $\theta_1 = \pi$.

The boundary conditions at the free end can be given as follows [46]

$$\begin{aligned} \pi R_1 \left[\bar{T}_s + \bar{S}_s \sin\alpha - \bar{N}_s \cos\alpha \right] &= 0 \\ \pi R_1 \left[\bar{M}_s - R_1 \left(\bar{N}_s \sin\alpha + \bar{S}_s \cos\alpha \right) \right] &= -M \\ \bar{u} \sec\alpha + \bar{v} + \bar{w} \tan\alpha &= 0 \\ \frac{1}{2} \frac{y}{L} \frac{\partial \bar{w}}{\partial y} + \frac{1}{R_1} \left[\bar{v} \sin\alpha - \bar{w} \cos\alpha \right] &= 0 \end{aligned} \tag{13.38a-d}$$

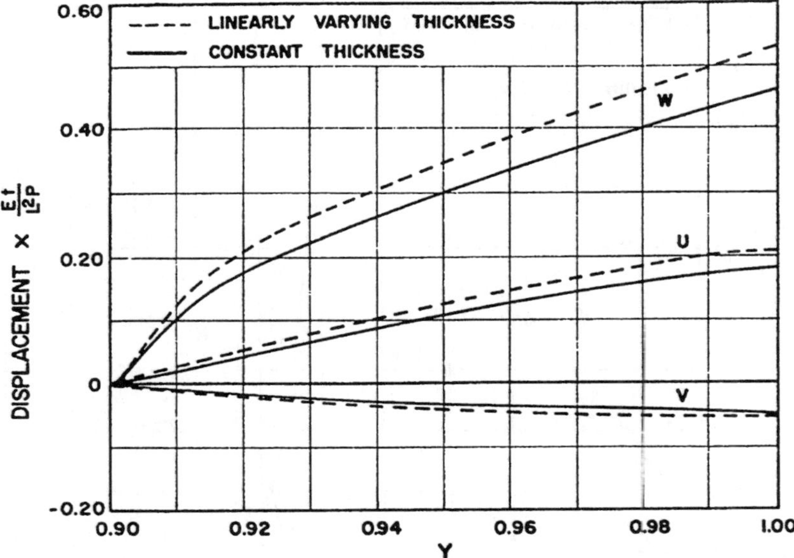

Fig. 13.3. Displacements u, v, and w $(n = 2)$

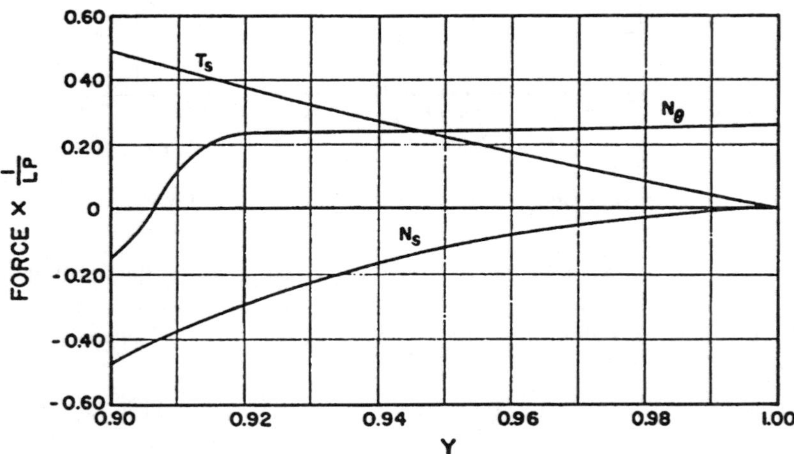

Fig. 13.4. Membrane forces N_θ, T_s, and N_s $(n = 2)$

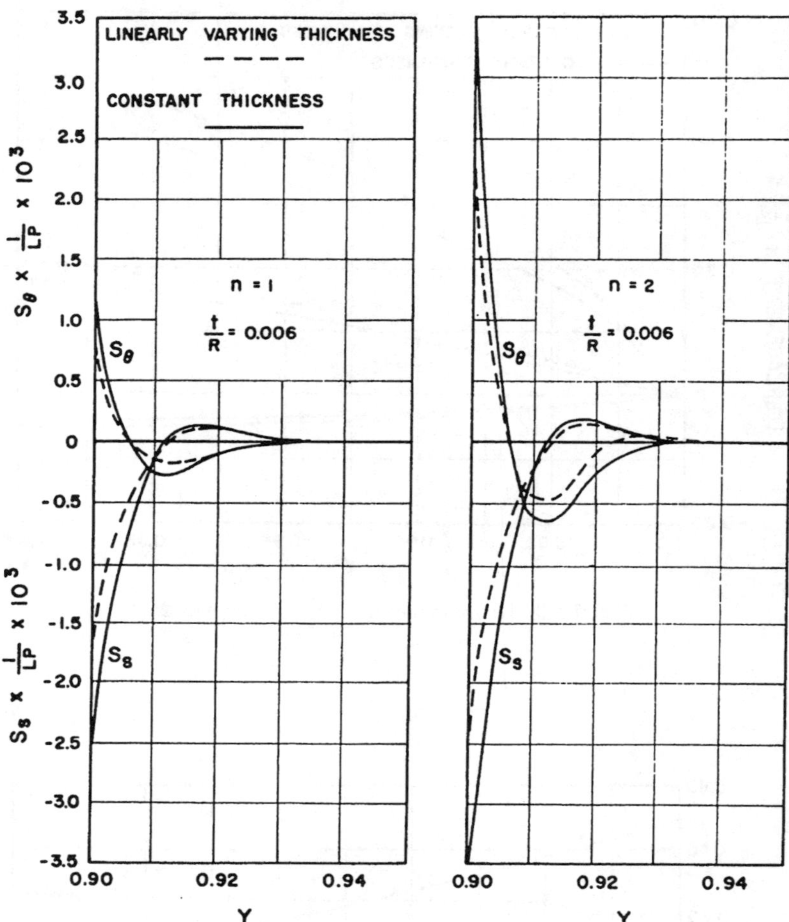

Fig. 13.5. Transverse shearing forces S_θ and S_s

where $R_1 = R\cos\alpha$ and a function with a bar at the top indicates that it is a function of y only. When the asymptotic solutions are used and the terms of the lowest order of $1/\rho$ are retained, the conditions (13.38) become, at $y = 1$, respectively,

$$\begin{gathered}\bar{N}^1_{s\theta} - \bar{N}^1_s \cos\alpha = 0 \\ \bar{N}^1_s = \left(M/\pi R_1^2\right)(1/\sin\alpha) \\ \bar{u}^1 \sec\alpha + \bar{v}^1 + \left(\bar{w}^1 + \bar{w}^{11}\right)\tan\alpha = 0 \\ \frac{\partial \bar{w}^{11}}{\partial y} = 0\end{gathered} \tag{13.39a-d}$$

The other four boundary conditions (13.5a-d) at the fixed end are

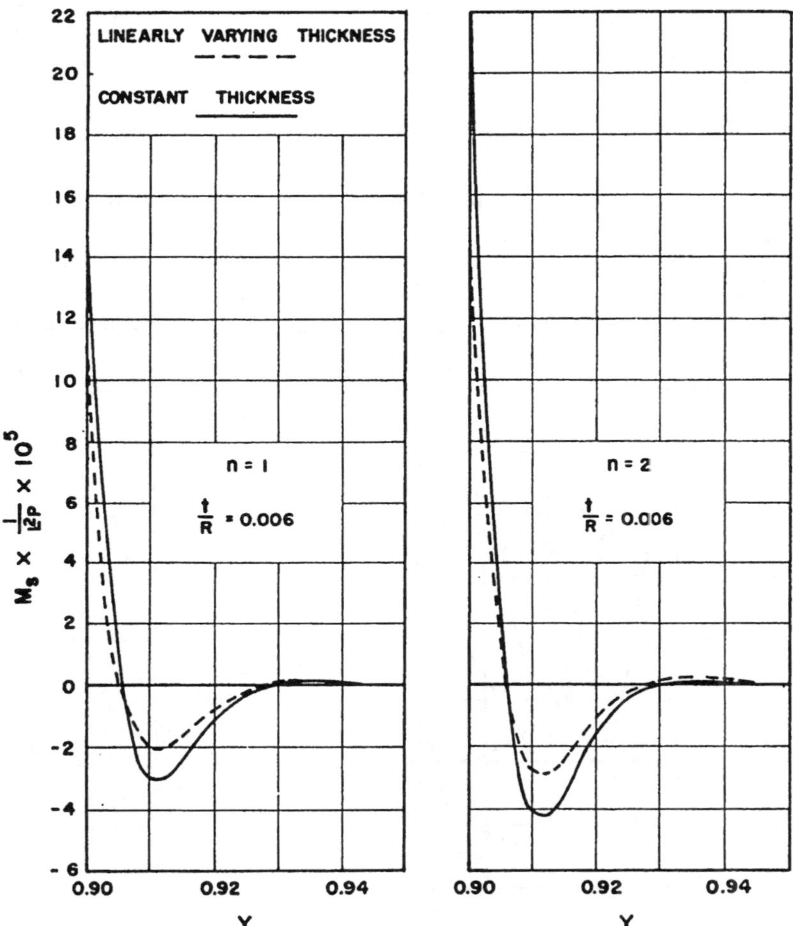

Fig. 13.6. Normal moments M_s

$$u^1 = v^1 = w^1 + w^{11} = \frac{\partial w^{11}}{\partial y} = 0 \quad \text{at} \quad y = (L_1/L)^{1/2} \tag{13.40}$$

The following material and geometrical constants are used:

$$\nu = 0.3 \qquad \frac{t}{R} = \frac{1}{40} \qquad \tan\alpha = \frac{4}{3} \qquad \frac{L_1}{L} = \frac{5}{8} \tag{13.41a-d}$$

Two sets of stress ratios, $\sigma_2/\sigma_{1\,\max}$ and $\sigma_m/\sigma_{1\,\max}$, were computed and are given in Fig. 13.7 along with the results of the Bessel function solutions given by Clark and Garibotti [12]. In Fig. 13.7

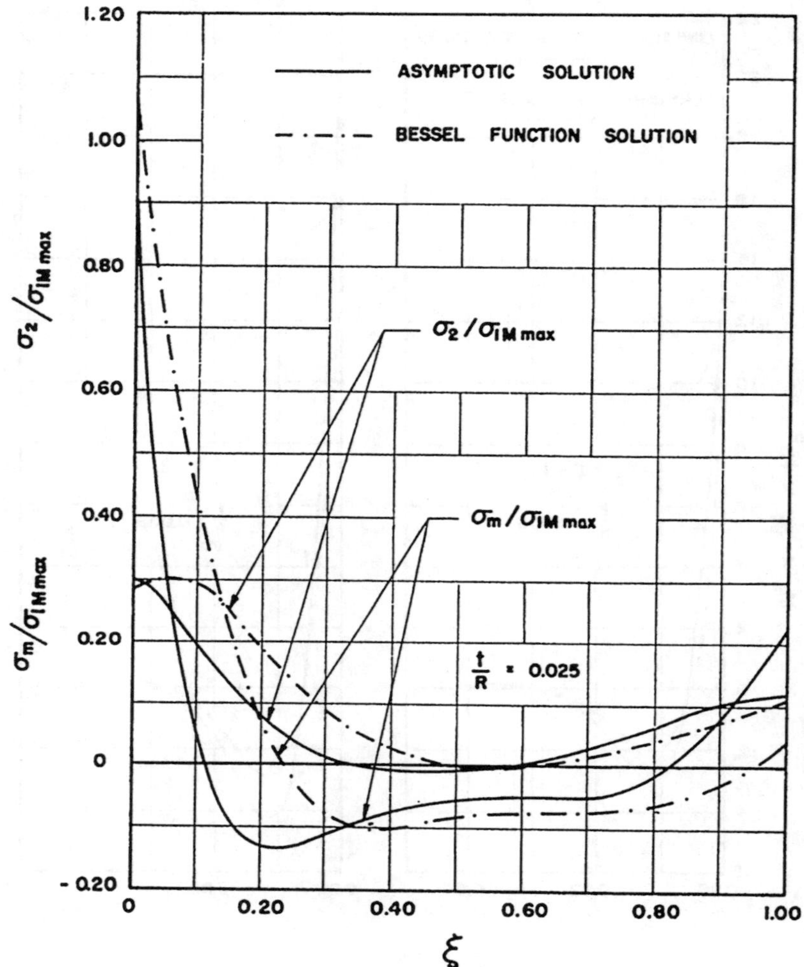

Fig. 13.7. Stresses for example 2

$$\sigma_1 M_{\max} = \left[\frac{N_s}{t}\right]_{\max} \qquad \sigma_2 = \frac{N_\theta}{t} \qquad \sigma_{\mathrm{m}} = \frac{6M_s}{t^2} \tag{13.42a-c}$$

and

$$\xi = \frac{y^2 - L_1/L}{1 - L_1/L} \tag{13.43}$$

The asymptotic solution obtained, like other asymptotic solutions, is simple in form, so that it is ready for direct applications and is approximate in nature. The accuracy depends on the ratio t/R; the smaller the ratio, the better the results will be. The shell used in this example is relatively thick for the application of the asymptotic solutions. Nevertheless, the results are still valuable for practical purposes since the agreement with the other known solutions is still fairly good.

13.6 Appendix

Let

$$\eta^{4/\gamma} = \sum_{n=0,1,\ldots}^{\infty} a_n p_n(\eta) \tag{13.A1}$$

where $p_n(\eta)$ are Legendre polynomials [16] and coefficients

$$a_n = \frac{2n+1}{2} \int_{-1}^{1} \eta^{4/\gamma} p_n(\eta)\, d\eta \tag{13.A2}$$

Then the series results in

$$\eta^{4/\gamma} = \frac{\gamma}{\gamma+4}\left[1 + \frac{1}{\gamma} Q_2(\eta) - \frac{1}{\gamma^2} Q_4(\eta) + \ldots\right] \tag{13.A3}$$

where

$$Q_2(\eta) = [20\gamma/(3\gamma+4)]\, p_2(\eta) \tag{13.A4a}$$

$$Q_4(\eta) = \{72(\gamma-2)\gamma/[(3\gamma+4)(5\gamma+4)]\}\, p_4(\eta) \tag{13.A4b}$$

Note that all $Q_n s\,(n = 2, 4, \ldots)$ are of the order $(1/\gamma)^0$.

The unknown function, W, may also be expanded in series

$$W = W_0(\eta) + (1/\gamma) W_1(\eta) + \left(1/\gamma^2\right) W_2(\eta) + \ldots \tag{13.A5}$$

Substitution of (13.A3 and 13.A5) into (13.21) results in a sequence of equations associated with various powers of $1/\gamma$. For the present study, the first term is all that wanted and thus

$$\eta^{4/\gamma} = \frac{\gamma}{\gamma+4} \tag{13.A6}$$

as given in (13.22).

Exercises

1. Verify (13.12) and check (13.13a).
2. Check (13.24 to 13.26) and try to obtain (13.33a).
3. Explain that from Figs. 13.1 to 13.6 for the constant thickness shell, the displacements are smaller, and moments and transverse shearing stresses are higher than those of the conical shell of linearly varying thickness.
4. Determine the stresses, N_s and N_θ, of the circular cone in Example 2 with the smaller end fixed and the other end free, subjected to the normal load defined in Example 1.

14

Membrane Vibrations of Conical Shells

The vibrations of conical shells are due to the inertial and external forces acting in the transverse, circumferential and meridional directions of the cone. For small vibrations, the system is linear, thus, the effects of these three forces may be treated separately. It may be due to the mathematical difficulty, the transverse vibration, so far, is solved by numerical methods only, such as Rayleigh-Ritz method, Galerkin procedure, power series, matrix method, and the finite element method. A comprehensive literature review on the conical shell vibrations is provided in [8].

Even for the simple problem of axial symmetry membrane transverse vibration, no complete analytic solutions were available [9]. In this chapter, analytic solutions for the membrane vibrations in the meridional and the circumferential directions are attempted. The latter may also be known as torsional vibration.

14.1 Basic Equations for Membrane Vibrations

The first three equations given by (12.1a-c) may be considered as equations of vibrations if D'Alembert's principle is applied, so that the inertial forces due to vibration are included in the external forces P_s, P_ϑ and P_z. They are acting in the directions indicated by the subscripts which are meridional, s, circumferential, θ, and normal, z. The three equations in these directions are, respectively,

$$\begin{aligned} \left(s\bar{N}_s\right)_{,s} + \left(\bar{N}_{\theta s,\theta}/\sin\alpha\right) - \bar{N}_\theta &= -sP_s \\ \left(s\bar{N}_{s\theta}\right)_{,s} + \left(\bar{N}_{\theta,\theta}/\sin\alpha\right) + \bar{N}_{\theta s} - \bar{Q}_\theta \cot\alpha &= -sP_\theta \\ \bar{N}_\theta \tan\alpha + \left(\bar{Q}_{\theta,\theta}/\sin\alpha\right) + \left(s\bar{Q}_s\right)_{,s} &= sP_z \end{aligned} \qquad (14.1\text{a-c})$$

in which the angle α is the complementary angle used in Chap. 12.

In (14.1), N_θ, N_s and Q_θ, Q_s are the normal and transverse shearing forces per unit length in the θ and s directions, respectively, and $N_{s\theta}(= N_{\theta s})$

is the in-plane shearing force per unit length. A prime followed by a subscript indicates a derivative with respect to the subscript.

14.2 Vibrations in the Meridional Direction

For a complete cone, let

$$\begin{aligned} \bar{U} &= hU(\eta)\cos(n\pi\theta/\theta_1)\exp(iqt) \\ \bar{V} &= hV(\eta)\sin(n\pi\theta/\theta_1)\exp(iqt) \\ \bar{W} &= hW(\eta)\sin(n\pi\theta/\theta_1)\exp(iqt) \end{aligned} \qquad (14.2\text{a-c})$$

$$\begin{aligned} \bar{N}_s &= EhN_s(\eta)\sin(n\pi\theta/\theta_1)\exp(iqt) \\ \bar{N}_\theta &= EhN_\theta(\eta)\sin(n\pi\theta/\theta_1)\exp(iqt) \\ \bar{N}_{s\theta} &= EhN_{s\theta}(\eta)\cos(n\pi\theta/\theta_1)\exp(iqt) \end{aligned} \qquad (14.3\text{a-c})$$

where $\bar{U}$, $\bar{V}$, and $\bar{W}$ are the displacements in the θ, s and z directions, respectively; h is thickness; t is time; E is the modulus of elasticity; q is the circular frequency; $\theta_1 = 2\pi$; and

$$\eta = s/L \qquad (14.4)$$

in which L is the distance measured from the apex to the large end of the cone.

The following force-displacement relations as given in (12.4a-c) are used for the present membrane theory

$$\begin{aligned} N_s &= \frac{k^2}{1-\nu^2}\frac{1}{\eta}\left[\eta V_{,\eta} + \nu\left(-pU + V + W\cot\alpha\right)\right] \\ N_\theta &= \frac{k^2}{1-\nu^2}\frac{1}{\eta}\left[-pU + V + W\cot\alpha + \nu\eta V_{,\eta}\right] \\ N_{s\theta} &= \frac{k^2}{1+\nu}\frac{1}{2\eta}\left[\eta U_{,\eta} - U + pV\right] \end{aligned} \qquad (14.5\text{a-c})$$

in which

$$p = (n\pi/\theta_1)/\sin\alpha \qquad k^2 = h/L \qquad (14.6\text{a,b})$$

and ν is Poisson's ratio.

Now consider the following case

$$P_s = \rho h^2 q^2 V + F(\eta, \theta, t) \quad P_\theta = 0 \quad P_z = 0 \qquad (14.7\text{a-c})$$

where ρ is the mass density and F is a given external force function. Because P_s acts along the surface of the cone, one may assume that

$$\bar{Q}_s = \bar{Q}_\theta = 0 \qquad (14.8\text{a,b})$$

Hence, (14.1c) calls for

$$\bar{N}_\theta = 0 \qquad (14.9)$$

Then (14.1a and 14.1b) are reduced, respectively, to

$$(\eta N_s)_{,\eta} - pN_{\theta s} = -\eta\rho h^2 q^2 V - \eta F \qquad (\eta N_{s\theta})_{,\eta} + N_{s\theta} = 0 \qquad (14.10,11)$$

14.3 Solutions for Free Vibrations

The solution of (14.11) is

$$N_{s\theta} = k^2 C_1/\eta^2 \tag{14.12}$$

in which C_1 is a constant of integration. The coefficient k^2 is introduced so that C_1 will be free from this magnitude parameter.

Because of (14.9), (14.5b) yields

$$W \cot \alpha = pU - V - \nu\eta V_{,\eta} \tag{14.13}$$

Substituting the above equation into (14.5a), one obtains

$$N_s = k^2 V_{,\eta} \tag{14.14}$$

Thus (14.10), for $F = 0$, assumes the form

$$\eta^2 V_{,\eta\eta} + \eta V_{,\eta} + \tau^2\eta^2 V = C_1 p/\eta \tag{14.15}$$

in which

$$\tau^2 = \rho L^2 q^2/E \tag{14.16}$$

where τ is the dimensionless frequency of free vibrations. The general solution of (14.15) is

$$V = C_2 J_0(x) + C_3 Y_0(x) + V_p \tag{14.17}$$

in which

$$x = \tau\eta \tag{14.18}$$

C_2 and C_3 are constants of integration, J_n and Y_n $(n = 0,\ 1)$ are the Bessel functions of the first and second kinds of the nth order, respectively, and V_p is the particular solution which may be obtained by the method of variation of parameters as

$$V_p = \tau p C_1 \left\{(1/x) + \left[J_0(x)\int_x Y_0(s)\,ds - Y_0(x)\int_x J_0(s)\,ds\right]\pi/2\right\} \tag{14.19}$$

The two integral functions involved in the last solution are available in [1,37].

Combining (14.5c) with (14.12) yields

$$xU_{,x} - U = 2(1+\nu)\tau C_1/x - pV \tag{14.20}$$

With a constant of integration, C_4, the solution of (14.20) is

$$\begin{aligned} U = C_4 x - (1+\nu)(\tau C_1/x) - px\Big\{C_2 \int_x \left[J_0(s)/s^2\right] ds \\ +C_3 \int_x \left[Y_0(s)/s^2\right] ds + C_1 \tau p \frac{1}{2x^2} - \pi T(x)\Big\} \end{aligned} \tag{14.21}$$

where

$$T(x) = \int_x \left[J_0(s)/s^2 \right] \left[\int_s Y_0(t)\, dt \right] ds - \int_x \left[Y_0(s)/s^2 \right] \left[\int_s J_0(t)\, dt \right] ds \tag{14.22}$$

The integral functions involved in the last equations may be carried out by first expressing the Bessel functions in the form of series, and then integrating them term by term. The results are in the forms of series of functions of x.

Substituting (14.17 and 14.21) into (14.13), one has

$$\begin{aligned} W \cot\alpha = {} & pC_1\tau \{ [(-2 + p^2/2)/x] - (\pi/2) [p^2 x T(x) + [J_0(x) \\ & - \nu x J_1(x)] \int_x Y_0(s)\, ds - [Y_0(x) - \nu x Y_1(x)] \int_x J_0(s)\, ds]] \} \\ & + C_2 \left[\left(p^2 - 1\right) J_0(x) - \left(p^2 - \nu\right) x J_1(x) + p^2 x \int_x J_0(s)\, ds \right] \\ & + C_3 \left[\left(p^2 - 1\right) Y_0(x) - \left(p^2 - \nu\right) x Y_1(x) + p^2 x \int_x Y_0(s)\, ds \right] + C_4 p x \end{aligned} \tag{14.23}$$

Finally, (14.14) becomes

$$\begin{aligned} N_s = {} & k^2 \tau \left\{ C_1 (\tau p \pi/2) \left[-J_1(x) \int_x Y_0(s)\, ds + Y_1(x) \int_x J_0(s)\, ds \right] \right\} \\ & - k^2 \tau \left[C_1 p \tau / x^2 + C_2 J_1(x) + C_3 Y_1(x) \right] \end{aligned} \tag{14.24}$$

The four boundary conditions needed for the determination of the four constants are

$$W = w \qquad N_{s\theta} = n_{s\theta} \tag{14.25a,b}$$

$$V = v \qquad N_s = n_s \tag{14.26a,b}$$

where w, v, $n_{s\theta}$ and n_s are prescribed values for the corresponding functions at the boundaries: $\eta = \beta$ and 1, where $\beta = L_1/L$ and L_1 is the distance measured from the apex to the small end of the cone; or $x = \tau\beta$ and τ. The solutions obtained above are exact in the frame work of linear membrane theory. Equations (14.12 and 14.11) indicate that if one end of a cone is free from the in-plane shearing force, $N_{s\theta}$, then there is no shearing stress throughout the cone. Thus for a cone with one end free from torsion, such as a cone with a free end or with one end simply supported but free to slide in the circumferential direction, one of the boundary conditions at that end is

$$N_{s\theta} = 0 \tag{14.27}$$

Then, in (14.12), $C_1 = 0$. This simplifies the solutions considerably. The numerical examples to be considered in the next section belong to this category.

14.4 Numerical Examples

In what follows, four kinds of cones satisfying (14.27) and

$$W = 0 \qquad \text{at } \eta = \beta \tag{14.28}$$

are considered. Therefore, at each end, one has two conditions to choose from: either $V = 0$ (fixed) or $N_S = 0$ (free).

14.4.1 Cones with the Small End Fixed (S-fixed) and Large End Free (L-free)

For such fixed-free cones, one requires

$$V = 0 \quad \text{at} \quad \eta = \beta \qquad N_s = 0 \quad \text{at} \quad \eta = 1 \tag{14.29,30}$$

With satisfaction of these two conditions, (14.17 and 14.24) yield, respectively,

$$C_2 J_0(\tau\beta) + C_3 Y_0(\tau\beta) = 0 \qquad C_2 J_1(\tau) + C_3 Y_1(\tau) = 0 \tag{14.31,32}$$

The condition of non-vanishing of the constants results in the following frequency equation of the system

$$J_0(\tau\beta)\, Y_1(\tau) - Y_0(\tau\beta)\, J_1(\tau) = 0 \tag{14.33}$$

The first five eigenvalues, τ, versus the end distance ratio, β, are depicted in Fig. 14.1 by the dash-and-dot lines. The displacement functions, V, $U' = U/k$, $W' = W(\cot\alpha)/P^2$, with $C_2 = 1$ and normalized N_s denoted as N'_s, are presented in Fig. 14.2 for $\beta = 0.5$.

For $\beta = 0$, one has a complete cone. Since $Y_0(0) = -\infty$, $C_3 = 0$. But $J_0(0) = 1$, thus the condition of (14.31) is fulfilled only if $C_2 = 0$, which yields a trivial solution. Physically, if the apex is fixed ($V = 0$, $W = 0$), there is a stress concentration which may be considered as a singular point of the system.

For large values of x,

$$J_0(x) = (2/\pi x)^{1/2} \cos(x - \pi/4) \quad Y_0(x) = (2/\pi x)^{1/2} \sin(x - \pi/4) \tag{14.34a,b}$$

$$J_1(x) = (2/\pi x)^{1/2} \cos(x - 3\pi/4) \quad Y_1(x) = (2/\pi x)^{1/2} \sin(x - 3\pi/4) \tag{14.34c,d}$$

Upon use of these asymptotic relations, (14.33) becomes

$$\cos\tau(1-\beta) = 0 \tag{14.35}$$

which yields the eigenvalues:

$$\tau = (2N-1)\,\pi/2\,(1-\beta) \qquad N = 1, 2, 3, \ldots \tag{14.36}$$

The curves of the above equations for $N = 1$, 2, 3, 4 and 5 are also depicted in Fig. 14.1 by the solid lines.

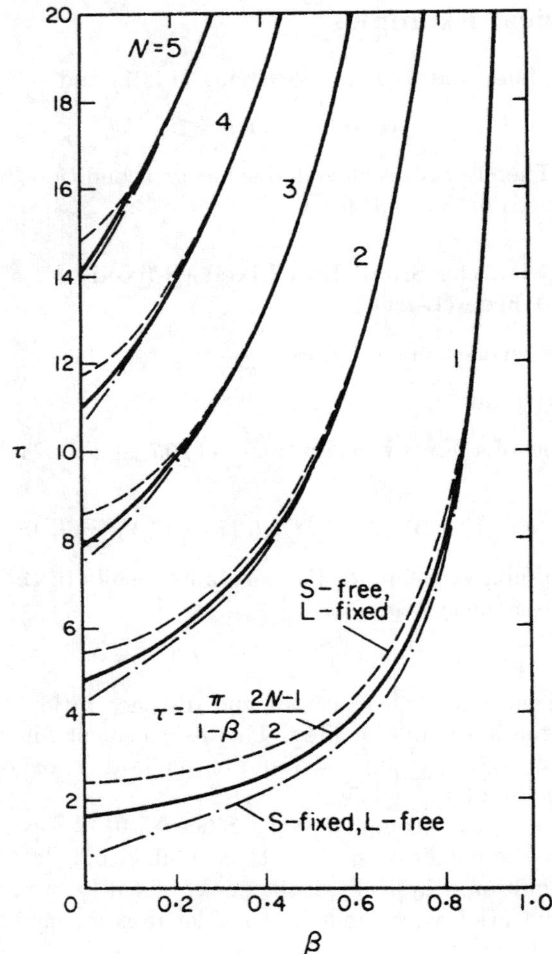

Fig. 14.1. First five frequencies of free-fixed and fixed-free cones

14.4.2 Cones with the Small End Free, (S-free) and Large End Fixed (L-fixed)

For such a cone, one requires

$$N_s = 0 \quad \text{at } \eta = \beta \quad V = 0 \quad \text{at } \eta = 1 \tag{14.37,38}$$

Thus one has

$$C_2 J_1 (\tau\beta) + C_3 Y_1 (\tau\beta) = 0 \quad C_2 J_0 (\tau) + C_3 Y_0 (\tau) = 0 \tag{14.39,40}$$

Then the frequency equation is

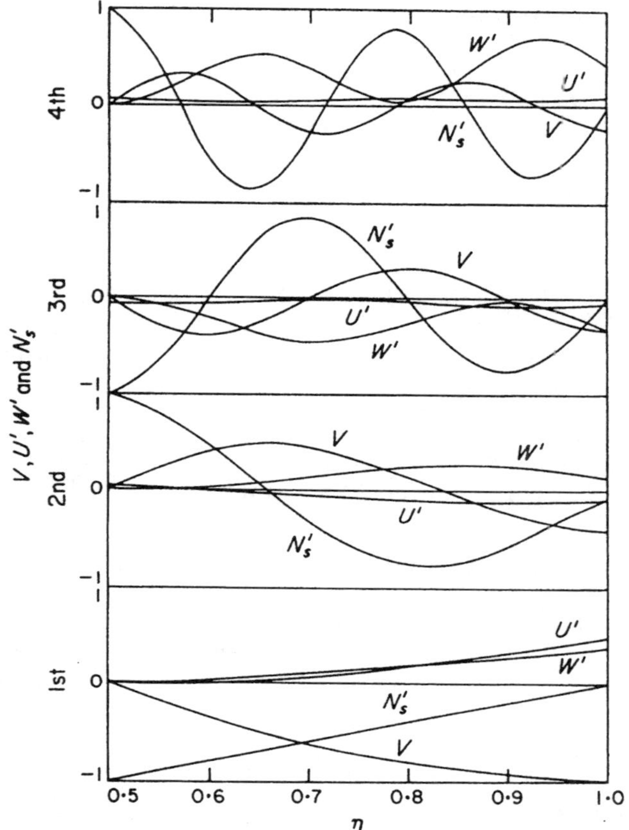

Fig. 14.2. Displacements V, U', W', and stress N_s of the first four modes of an S-fixed and L-free cone $\beta = 0.5$

$$J_0(\tau) Y_1(\tau\beta) - Y_0(\tau) J_1(\tau\beta) = 0 \tag{14.41}$$

which has the same asymptotic equation as (14.35). The numerical results of (14.41) are presented by the dashed lines in Fig. 14.1.

When $\beta = 0$, $J_1(0) = 0$, and $Y_1(0) = -\infty$, (14.40) is fulfilled for $C_2 \neq 0$ and $C_3 = 0$. For a complete cone with free apex, the eigenvalues, τ, are determined from the simple equation

$$J_0(\tau) = 0 \tag{14.42}$$

14.4.3 Cones with Both Ends Free (Free-free)

The two boundary conditions for such cones are

$$N_s = 0 \quad \text{at } \eta = \beta \text{ and } 1 \tag{14.43}$$

The frequency equation is

$$J_1(\tau) Y_1(\tau\beta) - Y_1(\tau) J_1(\tau\beta) = 0 \tag{14.44}$$

The asymptotic expression of (14.44) is

$$\sin \tau (\beta - 1) = 0 \tag{14.45}$$

The eigenvalues are

$$\tau = N\pi / (1 - \beta) \quad N = 1, 2, 3, \ldots \tag{14.46}$$

For complete cones, $\beta = 0$, non-trivial solutions exist. The eigenvalues, τ, may be determined from

$$J_1(\tau) = 0 \tag{14.47}$$

The numerical results of (14.44, 14.46 and 14.47) are depicted in Fig. 14.3. The eigenvalues of (14.42 and 14.47) are available in [37, 56].

14.4.4 Cones with Both Ends Fixed (Fixed-fixed)

These cones call for

$$V = 0 \quad \text{at} \quad \eta = \beta \text{ and } 1 \tag{14.48}$$

The frequency equation is

$$J_0(\tau) Y_0(\tau\beta) - Y_0(\tau) J_0(\tau\beta) = 0 \tag{14.49}$$

The first five frequencies are presented in Fig. 14.3 by the dash-dot curves. The asymptotic values are the same as in (14.45). For the same reason, for a fixed apex as discussed earlier, there are no non-trivial solutions for a complete cone of this type.

14.5 Forced and Transient Vibrations

The formal solutions for the forced and transient vibrations of the cones with $N_{s\theta} = 0$ are presented briefly herein by using the method of normal function expansions. Let the force function, defined in (14.7a), be given as

$$F = \left(h^2/L^2 E\right) \phi(\eta) \sin(n\pi\theta/\theta_1) \sin \Omega t \tag{14.50}$$

where ϕ is a prescribed function and Ω is a given frequency. The function, ϕ, may be expanded in the series of the normal functions $V_s(\eta)$ as

$$\phi(\eta) = \sum_{s=1.2\ldots} A_s V_s(\eta) \tag{14.51}$$

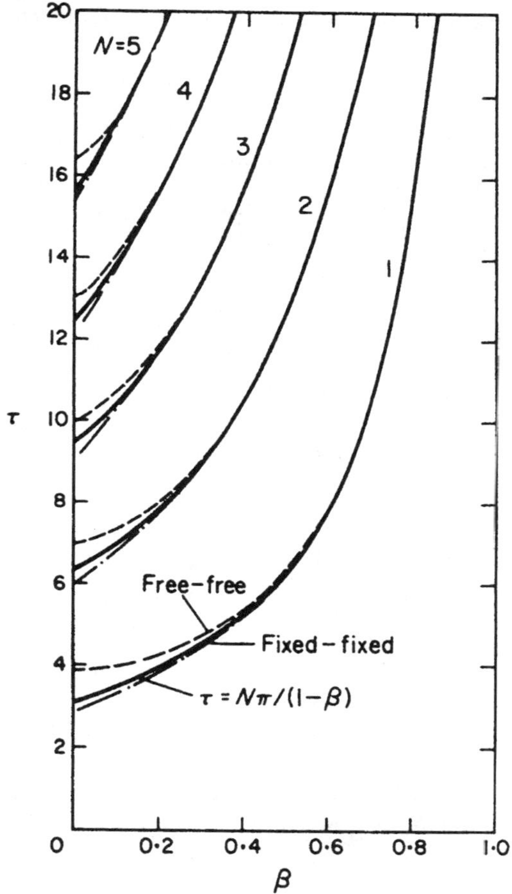

Fig. 14.3. First five frequencies of free-free and fixed-fixed cones

where the A_s are coefficients to be determined. The sth normal function satisfies

$$\eta V_{s,\eta\eta} + V_{s,\eta} = -\eta a^2 q_s^2 V_s \tag{14.52}$$

where

$$a^2 = \rho L^2/E \qquad q_s = \tau_s/a \tag{14.53a,b}$$

From (14.52), the orthogonality condition of the normal functions can be expressed as

$$\int_\beta^1 \eta V_r V_s d\eta = 0 \quad \text{for } r \neq s \tag{14.54}$$

Let

$$I = \int_\beta^1 \eta V_r^2 d\eta \tag{14.55}$$

Then the coefficients A_s in the series (14.51) may be determined as

$$A_s = \frac{1}{I} \int_{\beta}^{1} \eta \phi (\eta) V_s (\eta) d\eta \tag{14.56}$$

The motion equation for forced vibration is

$$\eta V_{,\eta\eta} + V_{,\eta} = \eta a^2 V_{,tt} - \eta \phi \sin \Omega t \tag{14.57}$$

with initial conditions

$$V (\eta, 0) = X (\eta) \qquad V_{,t} (\eta, 0) = Y (\eta) \tag{14.58a,b}$$

Now the formal complete solution, if $\Omega \neq q_s$, is

$$V (\eta, t) = \sum_{s=1,2,\dots} V_s (\eta) [B_s \cos q_s t + C_s \sin q_s t + D_s \sin \Omega t] \tag{14.59}$$

where

$$B_s = \frac{1}{I} \int_{\beta}^{1} \eta V_s (\eta) X (\eta) d\eta \tag{14.60}$$

$$C_s = \frac{1}{q_s} \left[\frac{1}{I} \int_{\beta}^{1} \eta V_s (\eta) Y (\eta) d\eta - \Omega D_s \right] \tag{14.61}$$

$$D_s = A_s / a^2 \left(q_s^2 - \Omega^2 \right) \quad \text{for } q_s \neq \Omega \tag{14.62}$$

On substitution of (14.59) into (14.20, 14.13 and 14.14), the corresponding solutions for U, W and N_s respectively, may be obtained.

14.6 Axisymmetric Torsional Vibrations

For torsional vibration, the inertial force acts in the circumferential direction, therefore, in (14.1)

$$P_s = 0 \quad P_\theta = -s\rho h \bar{U}_{,tt} \quad \text{and} \quad P_z = 0 \tag{14.63a-c}$$

Since $Q_s = Q_\theta = 0$, for axisymmetric case terms with differentiation with respective to θ vanish. Hence, (14.1a,c) are reduced to

$$\bar{N}_s = C/s \qquad \bar{N}_\theta = 0 \tag{14.64a,b}$$

in which C is a constant of integration. Equation (14.1b) becomes

$$\left(s\bar{N}_{s\theta} \right)_{,s} + \bar{N}_{\theta s} = s\rho h \bar{U}_{,tt} \tag{14.65}$$

$\bar{N}_s$ which is not involved in the rest of the formulations may, hence, be dropped hereafter. Let

$$\bar{N}_{s\theta} = EhN_{s\theta}(\eta)\exp(iqt) \tag{14.66}$$

$$\bar{U} = hU(\eta)\exp(iqt) \tag{14.67}$$

in which E is Young's modulus; q is the circular frequency; and $\eta = s/L$ where L is the distance measured from the apex to the large end of the cone as defined earlier. From (14.5c)

$$N_{s\theta} = \frac{k^2}{1+\nu}\frac{1}{2\eta}(\eta U_{,\eta} - U) \tag{14.68}$$

Equation (14.65) may be written as

$$\eta^2 U_{,\eta\eta} + \eta U_{,\eta} - \left(1 - \tau^2\eta^2\right)U = 0 \tag{14.69}$$

where

$$k^2 = h/L \quad \tau^2 = (Lq)^2\rho/G \quad G = E/\left[2(1+\nu)\right] \tag{14.70a-c}$$

in which G is known as torsional rigidity and ν is Poisson's ratio.

14.7 Solutions and Frequency Equations for Four Cases

The solution of (14.69) is

$$U = C_1 J_1(\tau\eta) + C_2 Y_1(\tau\eta) \tag{14.71}$$

in which C_1 and C_2 are constants of integration. J_n and $Y_n\,(n = 0,1)$, as defined before, are the Bessel functions of first and second kinds of the nth order, respectively. Substituting solution (14.71) into (14.68), one has

$$N_{s\theta} = \frac{k^2}{2(1+\nu)\eta}\frac{1}{\eta}\left\{C_1\left[\eta\tau J_0(\tau\eta) - 2J_1(\tau\eta)\right] + C_2\left[\eta\tau Y_0(\tau\eta) - 2Y_1(\tau\eta)\right]\right\} \tag{14.72}$$

The following four cases with different boundary conditions are considered:
Case (1). Small End Fixed (S-fixed) and Large End Free (L-free)

$$U(\beta) = 0 \qquad N_{s\theta}(1) = 0 \tag{14.73a,b}$$

Case (2). Small End Free (S-free) and Large End Fixed (L-fixed)

$$N_{s\theta}(\beta) = 0 \qquad U(1) = 0 \tag{14.74a,b}$$

Case (3). Both Ends Free (Free-free)

$$N_{s\theta}(\beta) = 0 \qquad N_{s\theta}(1) = 0 \tag{14.75a,b}$$

Case (4). Both Ends Fixed (Fixed-fixed)

$$U(\beta) = 0 \qquad U(1) = 0 \tag{14.76a,b}$$

in which $\beta = L_1/L = R_1/R$, L_1 is the distance measured from the apex to the small end and R_1 and R are the radii of the small and large ends, respectively. Thus $\beta = 0$ ($L_1 = 0$) is a complete cone and when $\beta = 1$ ($R_1 = R$), the cone becomes a ring.

The frequency equations for the four cases are, respectively,

$$J_1(\beta\tau)[\tau Y_0(\tau) - 2Y_1(\tau)] - Y_1(\beta\tau)[\tau J_0(\tau) - 2J_1(\tau)] = 0 \tag{14.77}$$

$$J_1(\tau)[\beta\tau Y_0(\beta\tau) - 2Y_1(\beta\tau)] - Y_1(\tau)[\beta\tau J_0(\beta\tau) - 2J_1(\beta\tau)] = 0 \tag{14.78}$$

$$\begin{aligned}[\beta\tau J_0(\beta\tau) - 2J_1(\beta\tau)][\tau Y_0(\tau) - 2Y_1(\tau)] \\ - [\beta\tau Y_0(\beta\tau) - 2Y_1(\beta\tau)][\tau J_0(\tau) - 2J_1(\tau)] = 0\end{aligned} \tag{14.79}$$

$$J_1(\beta\tau) Y_1(\tau) - Y_1(\beta\tau) J_1(\tau) = 0 \tag{14.80}$$

For large values of x

$$J_0(x) = -Y_0(x) = (2/\pi x)^{1/2}(\sin x + \cos x) \tag{14.81a}$$

$$J_1(x) = Y_1(x) = (2/\pi x)^{1/2}(\sin x - \cos x) \tag{14.81b}$$

Equations (14.77 to 14.80), by using these asymptotic relations, become, respectively,

$$\tan\Omega = \Omega/2(1-\beta) \tag{14.82}$$

$$\tan\Omega = -\beta\Omega/2(1-\beta) \tag{14.83}$$

$$\tan\Omega = 2\Omega(1-\beta)^2/\beta\left[\Omega^2 + 4(1-\beta)^2\right] \tag{14.84}$$

$$\sin\Omega = 0 \tag{14.85}$$

where

$$\Omega = (1-\beta)\tau = (L - L_1) q (\rho/G)^{1/2} \tag{14.86}$$

which is used as the frequency parameter, should not be confused with the Ω in (14.50). The first five frequencies of Ω versus the end distance ratio, β, are presented in Figs. 14.4 and 14.5 by solid and dash-and-dot curves for those obtained from the frequency equations, and dotted curves from the asymptotic equations accompany the respective curves.

For the limited case of $\beta = 0$, frequency (14.77 and 14.79) become

$$\tau J_0(\tau) - 2J_1(\tau) = 0 \tag{14.87}$$

and (14.78 and 14.80) are reduced to

$$J_1(\tau) = 0 \tag{14.88}$$

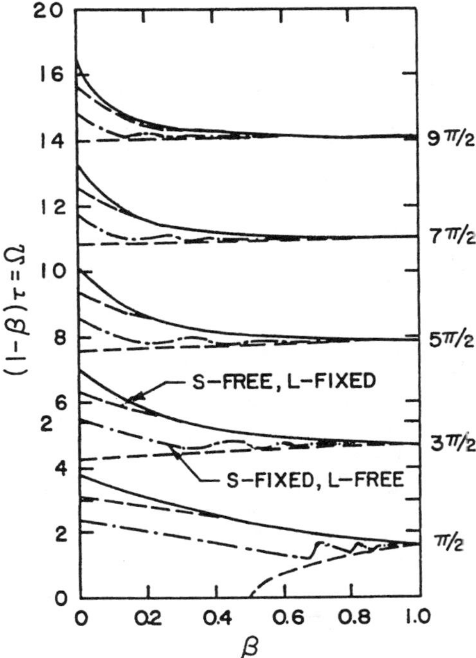

Fig. 14.4. First five frequencies of S-fixed, L-free and S-free, L-fixed cones

The results of (14.87), however, agree with that of (14.85) only for the high frequencies.

While $\beta = 1$, the limiting value of Ω may be obtained from the asymptotic equations. For S-fixed, L-free, and S-free, L-fixed, (14.82 and 14.83) become

$$\tan \Omega = \pm\infty \tag{14.89}$$

Hence

$$\Omega = \pi\,(2n-1)\,/2 \qquad n = 1, 2, 3, \ldots \tag{14.90}$$

For the other two cases, (14.84 and 14.85) are reduced to

$$\sin \Omega = 0 \tag{14.91}$$

Therefore,

$$\Omega = n\pi \qquad n = 1, 2, 3, \ldots \tag{14.92}$$

The asymptotic results of (14.90 and 14.92) are shown in Figs. 14.4 and 14.5, respectively, for $\beta = 1$.

Finally, it can be shown that the orthogonality condition for the nth normal function, U_n , has the following form:

$$\int_{\beta}^{1} \eta U_n U_m d\eta = 0 \quad \text{for} \quad n \neq m \tag{14.93}$$

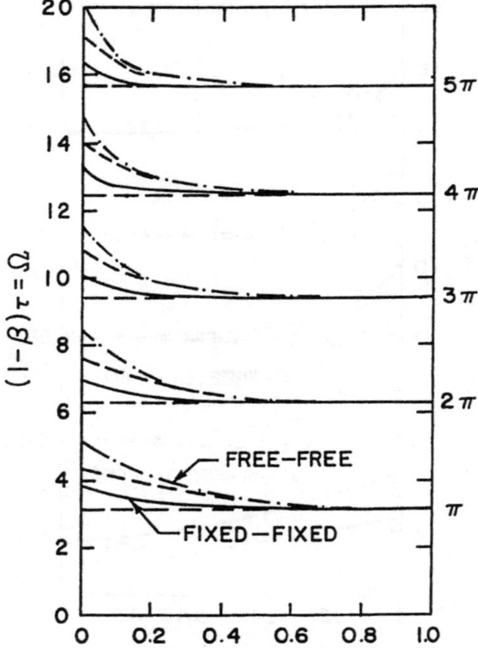

Fig. 14.5. First five frequencies of free-free and fixed-fixed cones

Exercises

1. Verify that (14.17 and 14.19) are the solutions of (14.15).
2. Construct a diagram by changing the coordinates in Fig. 14.1 for the first five frequencies for the free-fixed and fixed-free cones from τ vs. β to $\tau(1-\beta)$ vs. β.
3. Show the orthogonality (14.54) from (14.52).
4. Verify (14.72).

Appendix

Elements of Matrix $[\mathbf{A}_{ij}]$ of (4.7)

```
Y = DTAN(B)
Z = 1.0D0/DCOS(B)
T2 = T1**2/R
T3 = T1**2*Y/R
T4 = T1**2*Z/R
A5 = A4
T5 = T1*S2*Y
T6 = T1*S2*Z
T7 = T1*S4*Y
T8 = T1*S4*Z
DO 10 I = 1,24
DO 10 J = 1,24
10 A(I,J) = 0.0D0
A(1,1) = DSIN(T1) – DSINH(T1)
A(1,2) = DCOS(T1) – DCOSH(T1)
A(1,9) = –1.0D0
A(2,3) = DSIN(T2)
A(2,5) = 1.0D0
A(2,7) = 1.0D0
A(3,9) = 1.0D0
A(3,17) = DCOS(B)
A(3,19) = DCOS(B)
A(3,21) = –DSIN(B)
A(4,5) = 1.0D0
A(4,7) = 1.0D0
A(4,17) = –DSIN(B)
A(4,19) = –DSIN(B)
A(4,21) = –DCOS(B)
A(5,1) = DCOS(T1) – DCOSH(T1)
A(5,2) = –DSIN(T1) – DSINH(T1)
A(5,4) = –S2
A(5,6) = –S2
A(6,4) = S2
```

```
A(6,6) = S2
A(6,16) = –S4
A(6,18) = –S4
A(7,1) = –DSIN(T1) – DSINH(T1)
A(7,2) = –DCOS(T1) – COSH(T1)
A(7,5) = A2/(S2**2)
A(7,7) = –A2/(S2**2)
A(7,17) = A4/(S4**2)
A(7,19) = –A4/(S4**2)
A(8,1) = T1*(DCOS(T1)+DCOSH(T1))
A(8,2) = –T1*(DSIN(T1) – DSINH(T1))
A(8,8) = –A2*R
A(8,16) = (A4*T1*DCOS(B))/S4
A(8,18) = (–A4*T1*DCOS(B))/S4
A(8,20) = –A4*R*DSIN(B)
A(9,3) = DCOS(T2)
A(9,4) = (A2*T1)/(S2*R)
A(9,6) = (–A2*T1)/(S2*R)
A(9,16) = (A4*T1*DSIN(B))/(S4*R)
A(9,18) = (–A4*T1*DSIN(B))/(S4*R)
A(9,20) = A4*DCOS(B)
A(10,8) = DSIN(T3)
A(10,9) = DCOS(T3)
A(10,11) = 1.0D0
A(10,13) = 1.0D0
A(11,11) = 1.0D0
A(11,13) = 1.0D0
A(11,22) = (DSIN(T8) – DSIN(T8))*DCOS(B)
A(11,23) = (DCOS(T8) – DCOSH(T8))*DCOS(B)
A(11,24) = DSIN(T4)*DSIN(B)
A(12,4) = DSIN(T5)
A(12,5) = DCOS(T5)
A(12,6) = DSINH(T5)
A(12,7) = DCOSH(T5)
A(12,15) = –1.0D0
A(13,15) = 1.0D0
A(13,22) = –(DSIN(T8) – DSINH(T8))*DSIN(B)
A(13,23) = –(DCOS(T8) – DCOSH(T8))*DSIN(B)
A(13,24) = DSIN(T4)*DCOS(B)
A(14,4) = S2*DCOS(T5)
A(14,5) = S2*DSIN(T5)
A(14,6) = S2*DCOSH(T5)
A(14,7) = S2*DSINH(T5)
A(14,10) = –1.0D0
A(14,12) = –1.0D0
```

A(15,10) = 1.0D0
A(15,12) = 1.0D0
A(15,22) = –(DCOS(T8) – DCOSH(T8))*S4
A(15,23) = (DSIN(T8)+DSINH(T8))*S4
A(16,4) = (–A2/(S2**2))*DSIN(T5)
A(16,5) = (–A2/(S2**2))*DCOS(T5)
A(16,6) = (A2/(S2**2))*DSINH(T5)
A(16,7) = (A2/(S2**2))*DCOSH(T5)
A(16,11) = 1.0D0
A(16,13) = –1.0D0
A(16,22) = (–A4/(S4**2))*(DSIN(T8)+DSINH(T8))
A(16,23) = (–A4/(S4**2))*(DCOS(T8)+DCOSH(T8))
A(17,8) = A2*DCOS(T3)
A(17,9) = –A2*DSIN(T3)
A(17,10) = T1/R
A(17,12) = –T1/R
A(17,22) = (A4*(DCOS(T8)+DCOSH(T8))*DCOS(B))*T1/(S4*R)
A(17,23) = (–A4*(DSIN(T8)–DSINH(T8))*DCOS(B))*T1/(S4*R)
A(17,24) = A4*DCOS(T4)*DSIN(B)
A(18,4) = –(T1*A2*DCOS(T5))/(S2*R)
A(18,5) = (T1*A2*DSIN(T5))/(S2*R)
A(18,6) = (T1*A2*DCOSH(T5))/(S2*R)
A(18,7) = (T1*A2*DSINH(T5))/(S2*R)
A(18,14) = 1.0D0
A(18,22) = –(A4*T1*(DCOS(T8)+DCOSH(T8))*DSIN(B))/(R*S4)
A(18,23) = (A4*T1*(DSIN(T8)–DSINH(T8))*DSIN(B))/(R*S4)
A(18,24) = A4*DCOS(T4)*DCOS(B)
A(19,20) = DSIN(T4)
A(19,21) = DCOS(T4)
A(20,16) = DSIN(T8)
A(20,17) = DCOS(T8)
A(20,18) = DSINH(T8)
A(20,19) = DCOSH(T8)
A(21,16) = DCOS(T8)
A(21,17) = –DSIN(T8)
A(21,18) = DCOSH(T8)
A(21,19) = DSINH(T8)
A(22,10) = DSIN(T1)
A(22,11) = DCOS(T1)
A(22,12) = DSINH(T1)
A(22,13) = DCOSH(T1)
A(23,10) = DCOS(T1)
A(23,11) = –DSIN(T1)
A(23,12) = DCOSH(T1)

A(23,13) = DSINH(T1)
A(24,14) = DSIN(T2)

References

1. Abramowitz, M., and Stegun, I. J., *Hand Book of Mathematical Functions,* Dove Publication, New York, 1965.
2. Abrate, S., "Continuum Modeling of Latticed Structures," *The Shock and Vibration Digest*, 17(1), pp. 15–21, 1985.
3. Almroth, B. O., and Bushnell, D., "Computer Analysis of Various Shells of Revolution," *American Institute of Aeronautics Journal*, Vol. 5, No. 10, pp. 1848–1855, Oct., 1968.
4. Anderson, R. A., "Flexural Vibrations in Uniform Beams According to Timoshenko Theory," *Journal of Applied Mechanics* 20, pp. 504–540.1953.
5. Arbocz, J., "Buckling of Conical Shells under Axial Compression," NASA CR-1162, National Aeronautic and Space Administration. Washington, D.C., Sept., 1968.
6. Baruch, M., Harari, O., and Singer, J., "Low Buckling Loads of Axially Compressed Conical Shells," *Journal of Applied Mechanics*, Trans. ASME, Vol. 37, Series E, No. 2, p. 384, June, 1970.
7. Bicford, W. B., *Advanced Mechanics of Materials,* Prentice-Hall, Englewood Cliff, N.J., 1998.
8. Chang, C. H., "Vibrations of Conical Shells," *The Shock and Vibration Digest*, Vol. 13, No. 6, pp. 9–17, 1981.
9. Chang, C. H., "On Transverse Membrane Vibration of Conical Shells." *Journal of Applied Mechanics,* ASME, Vol. 51, pp. 213–214, March, 1984.
10. Chen, W. F., "Further Studies of an Inelastic Beam-Column Problem." *Journal of Structural Division,* ASCE, Vol. 97, No. ST2, Proc. Paper 7922, pp. 529–544, Feb., 1971.
11. Cheong Siat Moy, F., "Method of Analysis of Lateral Columns." *Journal of Structural Division,* ASCE, Vol. 100, No. ST5, Proc. Paper 10548, pp. 953–970, May, 1974.
12. Clark, R. A., and Garibotti, J. F., "Longitudinal Bending of a Conical Shell," *Proceedings of the 8th Midwestern Mechanics Conference, Developments in Mechanics*, Pt. 2, Vol. 2, pp. 113–130, Pergamon Press, New York, 1963.
13. Clough, R. W., and Jenschke, V. A., "The Effect of Diagonal Bracing on the Earthquake Performance of a Steel Frame Building", *Bulletin Seismological of America* 53, 1963.

14. Cowper, G. R., "The Shear Coefficient in Timoshenko's Beam Theory," *Journal of Applied Mechanics*, 33, pp. 335–340, 1966.
15. Craig, R. R., Jr., *Structural Dynamics*, John Wiley and Sons, New York, N.Y,, 1981.
16. Dettman, J. W., *Mathematical Method in Physics and Engineering,* Chapter 3, McGraw-Hill Book Company Inc., New York, 1962.
17. *Engineering and Design-Lock Gate, Engineering Manual,* EM 1110-2-2603, U.S. Army Corps of Engineering, Draft, Oct., 1974.
18. Falk, H., "Finite Element Analysis Packages for Personal Computers," *Mechanical Engineering* 107, pp. 54–71, 1985.
19. Flugge, W., *Stresses in Shells,* 3rd Printing, Springer-Verlag, Berlin, Germany, 1966.
20. Fung, Y. C., Wittrick, W. H., "A Boundary-Layer Phenomenon in the Large Deflection of Thin Plates," *Quarterly Journal of Mechanics and Applied Mathematics*, Vol. 8, p. 191, 1955.
21. Gere, J. M. and Weaver, W., *Analysis of Framed Structures*, Van Nostrand Reinhold, New York, N.Y. 1965.
22. Ghanaat, Y., "Study of X-braced Steel Frame Structures under Earthquake Simulation," Ph.D. Dissertation, The University of California, Berkeley, California. 1980.
23. Glefand, J. M. and Formin, S. V., *Calculus of Variations.* Prentice-Hall, Englewood Cliff, N.J., 1963.
24. Goel, S. C. and Hanson R. D., "Seismic Behavior of Multistory Braced Steel Frames," *Journal of the Structure Division*, ASCE, Vol. 100, pp. 79–95. 1974.
25. Homewood, R. H., Brine, A. C., and Johnson, A. E., "Experimental Investigation of the Buckling Stability of Monocoque Shells," *Experimental Mechanics*, pp. 88–96, Mar., 1961.
26. Hsu, Y. W., "The Shear Coefficient of Beams of Circular Cross Section," *Journal of Applied Mechanics* 42, pp. 226–228. 1975.
27. Huang, N. C., and Vahidi, B., "Snap-Through Buckling of Two Simple Structures," *International Journal of Non-Linear Mechanics*, Vol. 6, pp. 295–310, 1971.
28. Huang, T. C., "The Effect of Rotatory Inertia and of Shear Deformation on the Frequency and Normal Mode Equation of Uniform Beams with Simple End Conditions," *Journal of Applied Mechanics* 28, pp. 579–584. 1961.
29. Jensen, J. J., "On the Shear Coefficient in Timoshenko's Beam Theory," *Journal of Sound and Vibration*, 87, pp. 621–635. 1983.
30. Kollar, L., and Hededus, I., *Analysis and Design of Space Frames by Continuum Method.* Elsevier, New York, N.Y., 1985.
31. Kraus, H., *Thin Elastic Shells*, Wiley, Chapter 7, New York, N.Y. 1967.
32. Levinson, M., and Cooke, D. N., "On the Two Frequency Spectra of Timoshenko Beams," Journal *of Sound and Vibration* 84, pp. 319–326, 1982.
33. Lord Rayleigh, *The Theory of Sound,* Vol. 1, 2nd edition, Dove Publications, 2nd ed., New York..
34. Love, A. E. H., *A Treatise on the Mathematical Theory of Elasticity* , Dove Publication, New York, N.Y., 4th ed., 1944.
35. Mahony, J. J., "The Large Deflection of Thin Plates," *Quarterly Journal of Mechanics and Applied Mathematics*, Vol. 14, pp. 257–271, 1961.

36. Masur, E. F., and Chang C. H., "Development of Boundary Layers in Buckled Plates," *Journal of Engineering Mechanics Division*, Proceedings, ASCE, Vol. 90, No. EM2, pp. 33–50, Apr., 1964.
37. McLachlan, N. W., *Bessel Functions for Engineers*, the Clarendon Press, second edition, Oxford, 1955.
38. Mindlin, R. D., and Deresiewicz, H., "Timoshenko's Shear Coefficient for Flexural Vibrations of Beams," *Proceedings of the Second U.S. Congress of* Journal of Applied Mechanics 33, pp. 335–340. 1966. *Applied Mechanics*, pp. 175–178. 1955.
39. Novozhilov, V. V., *Thin Shell Theory*, 2nd revised ed., Chap. IV, P, Noordhoof Ltd., Groningen, The Netherlands, 1964.
40. Panovko, Y. G., and Gubanova, I. I., *Stability and Oscillations of Elastic Systems*, Chapter 2, translated from Russian by C. V. Larvick, Consultants Bureau, New York, N.Y., 1965.
41. Richards, T. H., and Leung, Y. T., "An Accurate Method in Structural Vibration Analysis," (see Appendix C). *Journal of Sound and Vibration,* 55, pp. 363–376. 1971.
42. Seide, P., "A Donnell Type Theory for Asymmetrical Bending and Buckling of Thin Conical Shells," *Journal of Applied Mechanics*, Vol. 24, No. 4, pp. 547–552, Dec., 1957.
43. Seide, P., "Axisymmetrical Buckling of Circular Cones under Axial Compression," *Journal of Applied Mechanics*, Vol. 23, No. 4, pp. 625–628, 1965.
44. Seide, P., "A Survey of Buckling Theory and Experiments of Circular Conical Shells of Constant Thickness," Report No. TDR-169 (3560–30) TN-1, Aerospace Corporation, Nov.,1962.
45. Singl, P., "Seismic Behavior of Braces and Braced Steel Frames," Report No. UMEE77R1, Department of Civil Engineering, The University of Michigan, Ann Arbor, Michigan.1977.
46. Thurston, G. A., "An Numerical Solution for Thin Conical Shell under Asymmetrical Loads, "*Proceedings of the* 4th *Midwestern Conference on Solid Mechanics,* pp. 171–194, 1959.
47. Timoshenko, S. P., Young, D. H., and Weaver, W., Jr. *Vibration Problem in Engineering.* John Wiley, Forth Edition, New York, N.Y., 1974.
48. Timoshenko, S. P., and Gere, J. M., *Theory of Elastic Stability*, McGraw-Hill Book Co., Inc., New York, N.Y., 1961.
49. Timoshenko. S. P., and Woinowsky-Krieger, S., *Theory of plates and Shells,* McGraw-Hill Book Co., Inc., New York, N.Y., 1959.
50. Tottenham, H., *Approximate Methods for Calculating Natural Frequencies and Dynamics Response of Elastic Systems*, Chapter 5, Lecture Notes in Engineering, edited by Brebbia, C. A., and Orszag, S. A., *Vibrations of Engineering Structures*, Springe-Verlag, 1985.
51. Trail-Nash, R. W., and Collar, A. R., "The Effect of Shear Flexibility and Rotatory Inertia on the Bending Vibrations of Beams," *Quarterly Journal of Mechanics and Applied Mathematics* 6, pp. 186–222, 1953.
52. *Uniform Building Code, International Conference of Building Officials*, Wittier, California, 1979 edition.
53. Vertes, G., *Structural Dynamics.* Elsevier, New York, N. Y., 1985.
54. Von Karman, T., and Kerr, A. D., "Instability of Spherical Shells Subjected to External Pressure," *Topics in Applied Mechanics*, D. Abir, F. Ollendorff, and

M. Reiner, eds., pp. 1–22, American Elsevier Publishing Co., New York, N.Y., 1965.

55. Wittrick, W. H., "A Nonlinear Theory for the Edge Stresses in Thin Shells," *International Journal of Engineering Science*, Vol. 2, p. 145, 1964.
56. Wylie, C. R., *Advanced Engineering Mathematics,* 4th ed., McGraw-Hill Book Co., Inc., New York, N.Y., 1975.

Index

Printing: Strauss GmbH, Mörlenbach
Binding: Schäffer, Grünstadt